# A Guide to Pharmaceutical Particulate Science

Timothy M. Crowder
Anthony J. Hickey
Margaret D. Louey
Norman Orr

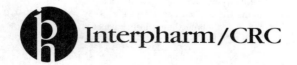



Interpharm/CRC

Boca Raton London New York Washington, D.C.

**Library of Congress Cataloging-in-Publication Data**

Crowder, Timothy M. [et al.]
A guide to pharmaceutical particulate science / Timothy M. Crowder, et al.
p. cm.
Includes bibliographical references and index.
ISBN 1-57491-142-2
1. Powders (Pharmacy) 2. Particles.
[DNLM: 1. Pharmaceutical Preparations—chemistry. 2. Powders—chemistry. 3. Drug Administration Routes. 4. Drug Compounding. QV 785 G946 2003] I. Crowder, Timothy M.
RS201.P8 G85 2003
615′.43—dc21 2002152626



Direct all inquiries to CRC Press LLC, 2000 N.W. Corporate Blvd., Boca Raton, Florida 33431.



**Visit the CRC Press Web site at www.crcpress.com**


International Standard Book Number 1-57491-142-2
Library of Congress Card Number 2002152626
Printed in the United States of America 1 2 3 4 5 6 7 8 9 0
Printed on acid-free paper

**Dr. Norman Orr**

*Powder technology is the foundation of dosage form design and particle engineering is the future of efficient, reproducible, and effective drug delivery.*

This was Norman's view of the justification for a book on pharmaceutical particulate science. His vision, enthusiasm, encouragement, and early contributions are its basis. This publication represents one small addition to Norman's list of achievements as a pharmaceutical scientist, educator, industrialist, colleague, family man, and friend. He is missed by all who knew him.

# The Authors

**Anthony J. Hickey, Ph.D., C. Biol. F.I.Biol.**, is Professor of Drug Delivery and Disposition and Biomedical Engineering and Mathematics at the University of North Carolina in Chapel Hill. In addition to his academic position, he is President and CEO of Cirrus Pharmaceuticals, Inc. and founder of Oriel Therapeutics, Inc.

Dr. Hickey received his Ph.D. in Pharmaceutical Sciences from the University of Aston in Birmingham, United Kingdom (1984), and then worked at the University of Kentucky before joining the faculty in Parmaceutics at the University of Illinois/Chicago. In 1990, he received the Young Investigator Award in Pharmaceutics and Pharmaceutical Technology from the American Association of Pharmaceutical Scientists. He accepted a faculty position in 1993 in the School of Pharmacy at the University of North Carolina. In 2000, he was elected a Fellow of the Institute of Biology of the United Kingdom.

Dr. Hickey has published over 100 papers and chapters in the pharmaceutical and biomedical literature and co-authored a text on pharmaceutical process engineering, as well as editing two volumes on the subject of pharmaceutical inhalation aerosols.

**Margaret D. Louey, Ph.D.**, graduated with a Bachelor of Pharmacy (Honors) and a Ph.D. in Pharmaceutics from the Victorian College of Pharmacy, Monash University (Australia). She completed a postdoctoral fellowship at the School of Pharmacy at the University of North Carolina/Chapel Hill. Her research has primarily involved particle interactions within dry powder aerosols and their formulations. Dr. Louey is a registered pharmacist with clinical experience in Australia and the United Kingdom.

**Timothy M. Crowder, Ph.D.**, graduated with a B.S. in Physics from Davidson College, Davidson, North Carolina and a Master's in Physics from North Carolina State University. He completed his Ph.D. in Biomedical Engineering from the University of North Carolina. Dr. Crowder's research concerns the physics of powder flow and its application to dispersion in dry powder inhalers. He is a founder and the Chief Technology Officer of Oriel Therapeutics.

# Contents

# Preface

Numerous texts focus on aspects of powder formation and behavior, notably the contributions of Rumpf, Rietema, and Allen. In addition, the pharmaceutical application of powder technology and particle science has long been recognized. Examples of books on this subject are those by Carstensen, Groves, and Barber. In recent years, some very thorough texts have addressed pharmaceutical powders and their properties. The present volume is intended to be a guide to some key principles and their practical applications. Numerous references throughout the text are listed alphabetically at the back of the book to direct the reader to sources for greater detail on elements of this exposition.

The idea for this book was conceived in 1993 by Norman Orr, then of SmithKline Beecham; Amy Davis, formerly of Interpharm Press; and me. Norman felt strongly that this need not be a comprehensive volume because many excellent texts were already available, such as those mentioned above. Instead, he suggested, and I agreed, that the importance of particle technology in the pharmaceutical sciences was understated, and that in particular the potential for error, misuse, or abuse of data should be addressed in a guide to the field. If this initiates some debate, I am sure Norman would have been very pleased.

In the intervening years, some quite dramatic developments have occurred. Most significantly, the driving force for the production of this book was lost. Norman Orr passed away prematurely after a relatively short illness in 1997. In addition, the level of understanding of particle properties and their importance for the performance of a pharmaceutical dosage form have increased substantially. This growth in knowledge was accompanied by a variety of technological advances in particle manufacture and characterization. Fortunately, these events occurred in time for Norman to see many of his aspirations realized. In the face of these changes, the incentive to produce such a volume waned. However, through the encouragement and support of Amy Davis and Tom Waters, of Interpharm Press, I continued to muse over the possibility of finishing this project. Finally, I was fortunate in that two enthusiastic and knowledgeable colleagues, Timm Crowder and Margaret Louey, joined me in the preparation of the manuscript that was the basis for this book. Largely through their persistence and hard work, this volume was completed.

Norman's major written contribution to the text is the introduction. It includes many of his comments, and their clarity and directness convey his commitment to this subject. His overriding view, expressed in early correspondence with me, was that "there is no current text that enables the pharmaceutical scientist to sensibly exploit the power of particulate science in the design, manufacture, and control of quality medicines." To the extent that this text helps in this endeavor, the credit can be given to Norman who, without doubt, was a visionary in this field. If it deviates from this goal, I accept the responsibility because Timm and Margaret have labored under my representations of Norman's objectives without the pleasure of having known him. I hope that you, the reader, find the final product of value in your day-to-day activities.

Anthony J. Hickey
Chapel Hill, NC, 2002

# 1

# Introduction and Overview

This volume will challenge current practice in the application of particle science in formulation design, manufacturing, and control of medicines. It aims to highlight appropriate means to exploit this science for the benefit of industry. The sections of this introduction address the relevant issues in pharmaceutical particulate science.

## Situation Analysis

The pharmaceutical reality is that in 90 percent of all medicines, the active ingredient is in the form of solid particles. Although the general population has a clear visual appreciation of particulate systems such as sand, gravel, stones, rocks, and boulders, and is capable of differentiating qualitatively among these, the ability to uniquely define particulate systems in a manner that suits all purposes is elusive.

Custom and practice in the pharmaceutical industry is such that particulate systems at best are poorly described and often are inadequately described to an extent that impacts the quality of the final product. Measurements need to relate to the state of the particles in the finished product and not just the raw material. The most applicable measurement may be one derived or interpolated from a number of techniques.

Particle size provides the basis for building quality into and optimizing the design of medicine, but it is seldom fully exploited. Raw materials are very inconsistent, and how this affects the activity of the medicines is unclear.

## Quality of Published Data on Particle Size in Pharmaceuticals

The following chapters review the literature with respect to the relationship of particle size to pharmaceutical parameters. The lack of rigor in the design of experiments is discussed, especially issues related to the following problems:

- Poorly prepared and even more poorly characterized size fractions
- The paucity of absolute measurements, including rigorous microscope counts or good photomicrographs
- The infrequent use of the complimentarity of particle sizing methods

## The Future of Crystal Engineering

Huge opportunities surround the engineering of surfaces and morphology of crystals. The implications for formulation design and batch-to-batch uniformity are immense. This should lead, in the first decade of the twenty-first century, to consistency in secondary manufacture. Ultimately, this approach should result in cheaper, well-designed, and higher-quality medicines.

## Chemistry and Pharmacy Regulatory Submissions

Increasingly, regulatory bodies such as the FDA need to be convinced that the formulation design is optimal, involving demands for particle size data of a caliber not seen previously. Pharmacopoeia will seek data of a quality at least equal to the American Society for Testing and Materials (ASTM) or the British Standards Institute.

## Misconceptions and Misunderstandings of Particulate Science

The pharmaceutical industry has an incomplete understanding of particulate science. Using tools that are not fully appreciated or understood may confuse or mislead the scientist or regulator.

## Powerful Methods Complementary to Particle Sizing

A mathematical interpretation of bulk properties such as flow of powders, coalescence of emulsions, and bleeding in ointments can be developed by using approaches in complexity analysis as well as stochastic and statistical phenomena. For instance, fractal analysis of powder flow can provide numerical values that contain a specific particle size component. Experimental investigations on the powder flow of pharmaceutical systems, including raw materials, excipients, intermediate granulations, and the lubricated granulation for encapsulation and tableting processes, should provide numerical values that relate to features of the component particles. These features may be simple, such as size, or more complex, such as a combination of size, shape, asperities, interparticle forces, and environment in interparticle spaces.

Comparison of such data with that derived from the application of such conventional techniques as sieving may enhance our understanding of the fundamental interactions leading to successful granulation. A broad view of how these data and their interpretation fit into pharmaceutical development embraces a wide range of concepts, from mathematical descriptors of particle shape and molecular probes of surface energetics to numerical definitions of bulk properties. Such methods enable revolutionary approaches to understanding processes such as disintegration, which in certain biological circumstances is a key precursor to dissolution.

The argument we are proposing is that particulate systems can be described, or evaluated, in three contrasting but complementary ways:

1. Direct phenomenological characterization (e.g., bulk flow, shear cell)
2. Knowledge of component particles, their surfaces, the microenvironment in which they exist, and theoretical models that support their interpretation
3. Mathematical interpretation of discontinuities resulting in the observed phenomena (e.g., coalescence of emulsions, avalanching powder flow, bleeding in ointment, and disintegration of tablets, or the sudden disruption of the tablet from the bulk). Such knowledge may translate into an improved product.

## Integration of Particle Sizing Methods with Their In-Use Situation

Powder science may be thought of as the understanding of particulate materials at a molecular level, their behavior as individual particles, and the physical and mechanical properties of collections of particles under defined conditions or in specific processes. Formulation can be defined as the transformation of a new chemical entity (NCE) into a medicine that is convenient to manufacture, distribute, and use, such that on administration the NCE is delivered in a predictable qualified and desirable manner. For the majority of medicines from the process of discovery/isolation of the NCE through production as a medicine until dissolution (e.g., in the gastrointestinal tract, or GIT), the NCE is in particulate form.

The nature of this particulate form (morphology, chemistry of the surface, and physical size) varies, depending upon its place in the transformation process. At any given time, the form of a particle may be critical to key parameters. For example, during secondary production, the shape, particle size, and surface energetics may be critical for successful granulation and subsequent compression; the solid-state stability of a capsule or tablet may well depend upon size morphology and excipient drug particle adjacencies; and the content uniformity of potent drugs depends upon both the mass of the individual particles and the extent to which these individual particles are adequately dispersed and not agglomerated.

A desirable feature of any particulate sizing program for a given NCE in a dosage form would be to monitor the size and characteristics of that particle during the entire process, from isolation at primary manufacture through to disintegration in the GIT. This, then, is the extent of the challenge. In some circumstances, this challenge is almost met, but in the vast majority of products on the market this is not the case by default, and formulation success and final quality of the product are not guaranteed. Consequently, it may be possible to consider a measure or expression for the particle size of a powder in the absence of knowledge of its intended use.

## Adjacencies and Interactions Between Particles

Particle–particle interactions do occur, and each particle must be considered in the context of its environment. Adjacencies and interactions between particles fall into two categories, advantageous and disadvantageous. The type of interactions being sought will depend upon the aspect of the design or process being considered.

As an example, when considering content uniformity and dissolution, the decision may be that particles should be <10 μm in size. These particles must be free flowing and must disperse and mix with excipients readily. If the mix is used in a direct compression process, it would be desirable that strong interaction occur between the excipient and drug but not between adjacent excipient particles (e.g., magnesium stearate when used as an excipient in tablet compression).

Technology exists to specifically assess the extent, at any processing stage, of particle adjacencies and interactions. Each of the steps is concerned with either breaking particle–particle interactions, or encouraging the formation of new particle–particle interactions, or both processes (i.e., mixing, granulation, tableting, disintegration).

## Lessons from Other Industries

Other industries have studied the behavior of particulates closely. For example, the efficiency of release of energy from the combustion of coal depends to large extent on the nature of the particles being employed. The performance of paints and polymers that consist of particulate material has been correlated with the characteristics of these particles. The modern printer utilizes a sophisticated method of depositing toner powders with unique physicochemical properties suitable for producing printed matter. These examples and others indicate the wealth of experience that can be tapped for application to pharmaceutical products.

## Pharmaceutical Particulate Science as a Core Educational Requirement

Currently, particulate science for pharmaceutical systems is taught in engineering and pharmacy schools on an ad hoc basis. A convincing argument can be made for collecting all the appropriate information into a core discipline of pharmaceutical particulate science for instruction to those students intending to work in the pharmaceutical industry.

## Particle Characterization and Its Impact on Regulatory Submissions from Formulation through Toxicology and Efficacy

The interface between the dosage form and the organism is the basis for the pharmacological or toxicological effects observed. The absolute amount of a drug or xenobiotic delivered by any route of administration and its rate of release, which will influence the bioavailability, must be considered since efficacy and toxicity are dose related. The nature of the particulate components of the drug delivery system will ultimately dictate the rate of release and instantaneous dose. The crystal structure, presence of polymorphs, and degree of subdivision all contribute to the solubility and dissolution rate of drugs. In general terms, these features influence the circulating concentrations of the agent, bringing it to therapeutic or toxic levels. However, in certain circumstances, specific organs may experience higher or lower concentrations of the drug than circulating levels based on intentional or unintentional targeting. Inhalation aerosol delivery, for example, deliberately targets the lungs and achieves locally therapeutic doses while minimizing systemic side effects. Oral delivery of drugs results in absorption to a blood supply, which is delivered to the liver by the hepatic portal vein before distribution to the rest of the body. This represents unintentional targeting of an organ, which may result in local advantageous or deleterious effects.

Minimally, a full characterization of the particulate nature of the product is required to ensure reproducibility of dose delivery and overall quality. If the particle size and distribution are thought to directly impact on the efficacy of a product, as is the case with inhalation aerosols, then this becomes a determinant of regulatory scrutiny.

## Conclusion

The role of particulate science in the preparation of medicines is ubiquitous. The solid state is the dominant means of presentation of a drug. Once molecules have crystallized from solution or solidified from a melt, the individual form and degree of subdivision of the population must be characterized. These features directly relate to performance of the dosage form and ultimately the way in which molecules present at the site of absorption, action, and the target receptor. The impact of the particulate nature of components of the dosage form on therapeutic effect is of the utmost importance and is considered in the following chapters. For convenience, the sequence outlined in this introduction has been divided into the following topics: particulate material, its form and production (Chapter 2); sampling from bodies of powder (Chapter 3); particle size descriptors and statistics (Chapter 4); behavior of particles (Chapter 5); instrumental analysis (Chapter 6); particle size measurement and synergy of adopted techniques (Chapter 7); physical behavior of a powder (Chapter 8); and in vitro and in vivo performance criteria (Chapter 9).

Figure 1.1 illustrates the issues that must be considered without proposing a relationship between them. The general conclusion to the book draws these components of pharmaceutical particulate science into a single concise description of their integration in the context of product development.

**Figure 1.1** Important factors in pharmaceutical particulate science.

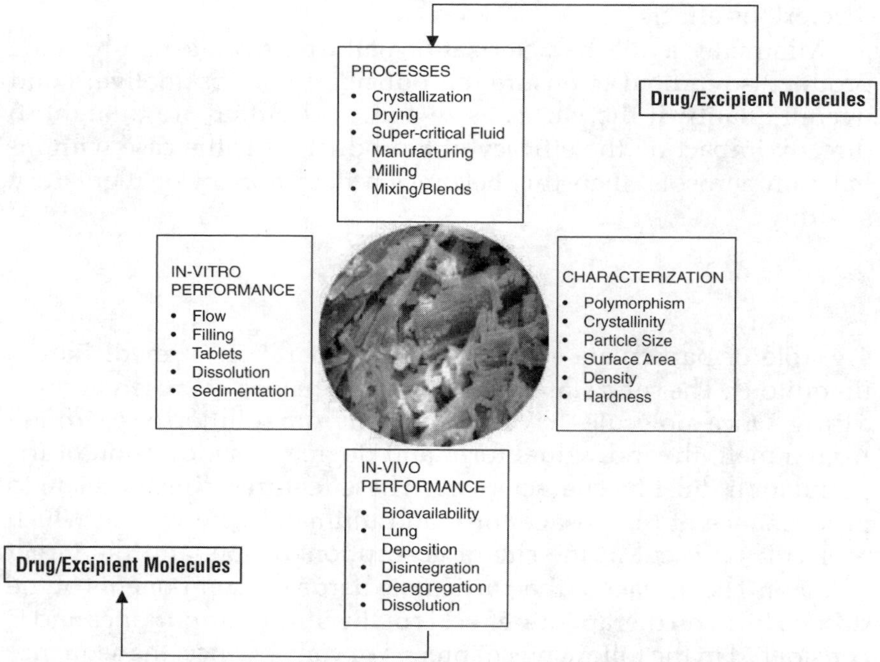

# 2

# Particulate Systems: Manufacture and Characterization

As an introduction to the methods of particle production, a brief review of the fundamentals of the states of matter and crystal systems may be helpful.

## States of Matter

The three states of matter—gas, liquid, and solid—each represent a different degree of molecular mobility. Gibbs first described the nature of states of matter according to the principles of thermodynamics (Rukeyser 1992; Gibbs 1993). Despite developments in the fields of quantum physics and chemistry, Gibbs' observations remain a valid description of the nature of matter.

Gas molecules are in constant vigorous, random motion according to the classical ideal gas theory of Bernoulli. Consequently, they take the shape of the container, are readily compressed, and exhibit low viscosity.

Liquids exhibit restricted random motion; the volume occupied is limited by their condensed nature resulting from intermolecular forces. Thus, liquids take the shape of a portion of the container. The liquid properties of water are within everyone's daily experience. However, to illustrate the extreme, other commonly experienced substances such as wax, pitch, and glass are highly viscous liquids, and not, as they appear to be, solids.

The last state of matter is highly constrained and is the result of a variety of forces. Solids are of great significance in the way we experience the world and are a dominant theme in the preparation of pharmaceuticals.

In broad terms, solids may be found in amorphous or crystalline states (Mullin 1993; Glusker, Lewis, and Rossi 1994). Crystallinity involves the regular arrangement of molecules or atoms into a fixed, rigid, three-dimensional pattern or lattice. Many amorphous materials exhibit some degree of crystallinity. The term "crystalline" is reserved for materials that exhibit a high degreee of internal regularity resulting in development of definite external crystal faces.

Truly amorphous solids are similar to gases and liquids in that their physical properties when measured in any direction are the same. In consequence, they are regarded as isotropic substances. Most crystals are anisotropic in that their mechanical, electrical, magnetic, and optical properties vary according to the direction of measurement. Cubic crystals are a notable exception because their highly symmetrical internal arrangement renders them optically isotropic. Anisotropy is most readily detected by refractive index measurement. Polarizing lenses may be employed to study anisotropy optically.

Liquid crystals are isotropic materials that can be induced to exhibit anisotropic behavior. Bose's swarm theory explains mesophase formation in liquid crystals by the formation of ordered regions in an otherwise randomly arranged "liquid." Molecules in these systems "swarm" into alignment. This phenomenon can be classified into smectic (soap-like), nematic (thread-like), and cholesteric. The smectic mesophase is characterized by flow of liquids in thin layers over each other. Cholesteric mesophases exhibit strong optical activity. Systems consisting of organic, often aromatic, elongated molecules form anisotropic liquids.

## Crystalline Solids

Crystals comprise a rigid lattice of molecules, atoms, or ions. The regularity of the internal structure of this solid body results in the crystal having characteristic shape (Jones and March 1973a, 1973b). The growth of most natural crystals has been restricted in one or more directions, resulting in exaggerated growth in other directions and giving rise to the so-called crystal habit (Tiwary 2001; Umprayn et al. 2001; Bennema 1992; Garekani 2001).

The apparent order exhibited by crystalline solids was once thought impossible to classify. In the late eighteenth century, however, Haüy, building on the observations of Steno from the prior century, proposed the law of constant interfacial angles, which stated that angles between corresponding faces of all crystals of a given substance are constant. Therefore, crystals may vary in size and development of various faces but the interfacial angles do not vary. A single substance can crystallize in more than one structural arrangement or form, known as polymorphism, an anomaly with regard to Haüy's law. However, the law does apply to each form, or polymorph. Modern techniques of X-ray crytallography enable lattice dimensions and interfacial angles to be measured, by the application of Bragg's law, with high precision on milligram quantities of crystalline powders (Nuffield 1966; Llacer et al. 2001; Darcy and Buckton 1998; Brittain 2001).

## Crystal Symmetry

Classification of crystals can be considered in terms of the apparent order conveyed visually. This can be described in terms of the symmetries exhibited, either about a point (center), a line (axis), or a plane. A cube is a simple example of such classification, having a single point, 13 axes, and 9 planes of symmetry, totaling 23 elements of symmetry, as shown in Figure 2.1. An octahedron exhibits the same quantity or elements of symmetry as a cube. Therefore, these polygons are related despite their apparent difference in appearance. The octahedron can be transformed into a cube by passing through the intermediate forms of truncated cube and octahedron and cubo-octahedron. These represent 3 of the 13 Archimedean semiregular

**Figure 2.1** Classification of (a) point (center), lines (axes), and (b) planes of symmetry for a cube.

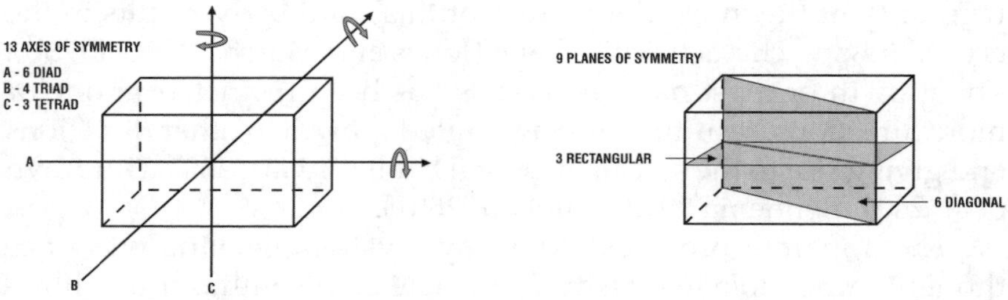


solids, which are called combination forms. Crystals frequently exhibit combination forms. The tetrahedron is also related to the cube and octahedron. These three forms belong to the five regular solids of geometry. The other two forms, the regular dodecahedron and icosohedron, do not occur in the crystalline state. However, the rhombic dodecahedron is frequently found in nature, particularly in garnet crystals.

## Euler's Relationship

Euler's relationship is particularly useful for calculating the number of faces (F), edges (E), and corners (C, vertices) of any polyhedron according to the expression

$$E = F + C - 2 \tag{2.1}$$

Intriguingly, Gibbs' phase rule, which describes the states of matter, appears to follow a similar expression in relating C, the number of components; F, the number of degrees of freedom; and P, the number of phases:

$$F = C - P + 2 \tag{2.2}$$

When rearranged, Gibbs' phase rule takes the same form as Euler's relationship:

**Figure 2.2** (a) Rotary inversion symmetry and (b) cubic geometry illustrating Miller indices.

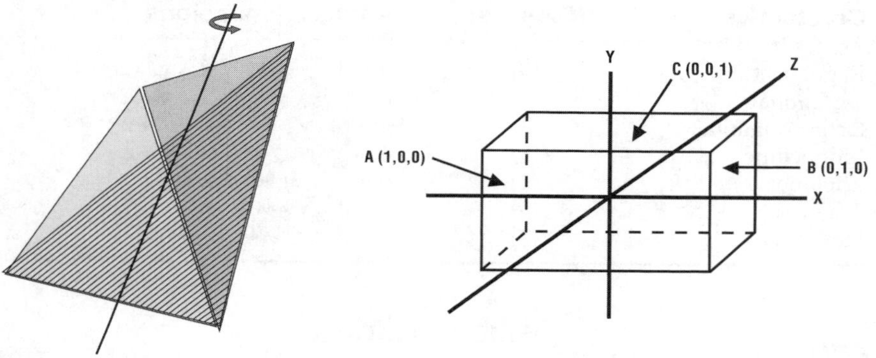


$$C = F + P - 2 \tag{2.3}$$

A fourth element of symmetry is known as compound or alternating symmetry, or, alternatively, symmetry about a rotation-reflection axis or axis of rotatory inversion. This type of symmetry comes about by performing two operations: rotation about an axis and reflection in a plane at right angles to the axis, as shown in Figure 2.2a. This is also called inversion about the center.

## Crystal Systems

Only 32 combinations of the elements of symmetry are possible. These are the point groups, or classes. All but a few of these classes have been observed in crystalline bodies. These classes are grouped into seven systems based on three dimensions (x, y, and z) and angles of faces ($\alpha$, $\beta$, and $\gamma$). Hexagonal crystals are unique because they have six edges in two dimensions rather than four as for the other crystal systems. Hexagonal crystals are described by invoking a third dimension in two-dimensional space with a third angle ($\mu$). Table 2.1 describes the crystal systems.

**Table 2.1** Crystal Systems, the Number of Subsidiary Classes, and Crystal Structures (Dimensions x, y, z, and Angles $\alpha$, $\beta$, $\gamma$, $\mu$)

| Crystal System | (Classes) | Angles/Dimensions |
|---|---|---|
| Regular | (5) | $\alpha = \beta = \gamma = 90°$ / $x = y = z$ |
| Tetragonal | (7) | $\alpha = \beta = \gamma = 90°$ / $x = y \neq z$ |
| Orthorhombic | (3) | $\alpha = \beta = \gamma = 90°$ / $x \neq y \neq z$ |
| Monoclinic | (3) | $\alpha = \beta = 90° \neq \gamma$ / $x \neq y \neq z$ |
| Triclinic | (2) | $\alpha \neq \beta \neq \gamma \neq 90°$ / $x \neq y \neq z$ |
| Trigonal | (5) | $\alpha = \beta = \gamma \neq 90°$ / $x = y = z$ |
| Hexagonal | (7) | $\alpha = \beta = \mu = 60°, \gamma = 90°$ |

## Miller Indices

An alternative visual reference technique known as Miller indices involves using a three-dimensional Cartesian coordinate system to indicate the position of crystal faces in space. In this technique, all of the faces of a crystal can be described in terms of their axial intercepts, as illustrated in Figure 2.2b.

## Space Lattices

Hooke and Haüy concluded that all crystals were built up of a large number of minute units, each shaped like the larger crystal. Consequently, a space lattice is a regular arrangement of points in three dimensions. Each of the seven crystal systems identified by elements of symmetry can be classified into 14 Bravais lattices, as shown in Table 2.2. Although these are the basic lattices, interpenetration can occur in actual crystals; this interpenetration can potentially give rise to 230 possible combinations. An alternative Bravais-Donnay-Harker principle considers space groups rather than lattice types. Generally, crystals cleave along lattice planes.

## Solid State Bonding

Four types of crystalline solid may be specified: ionic, covalent, molecular, and metallic. Some materials are intermediate between these types.

**Table 2.2** Fourteen Bravais Lattices and Their Equivalent Symmetries and Crystal Systems

| Type of Symmetry | Lattice | Crystal System |
|---|---|---|
| Cubic | Cube<br>Body centered<br>Face centered | Regular |
| Tetragonal | Square prism<br>Body centered | Tetragonal |
| Orthorhombic | Rectangular prism<br>Body centered<br>Rhombic prism<br>Body centered | Orthorhombic |
| Monoclinic | Monoclinic parallelepiped<br>Clinorhombic prism | Monoclinic |
| Triclinic | Triclinic parallelepiped | Triclinic |
| Rhomboidal | Rhombohedron | Trigonal |
| Hexagonal | Hexagonal prism | Hexagonal |

Ionic crystals (e.g., sodium chloride) consist of charged ions held in place in the lattice by electrostatic forces. Each ion is separated from oppositely charged ions by regions of negligible electron density.

Covalent crystals (e.g., diamond) consist of constituent atoms, which do not carry effective charges. A framework of covalent bonds, through which their outer electrons are shared, connects these atoms.

Molecular crystals (e.g., organic compounds) are discrete molecules held together by weak attractive forces (π-bonds, hydrogen-bonds).

Metallic crystals (e.g., copper) comprise ordered arrays of identical cations. The constituent atoms share their outer electrons, which are free to move through the crystal and confer "metallic" properties on the solid.

Pharmaceutical products are largely limited to ionic and molecular crystals. The ionic crystals are associated with additives to the products; the drug substance itself is usually a molecular crystal. However, exceptions include sugars (e.g., lactose), which may be used as excipients, and salts (e.g., potassium chloride), which have physiological activity.

## Isomorphs and Polymorphs

Polymorphs are different crystalline forms of the same chemical entity, which may exhibit different physicochemical properties. Different polymorphic forms or changes in solvation may be detected by using bulk analytical methods, including thermal analysis such as differential scanning calorimetry (DSC) and thermogravimetric analysis (TGA) (Turi, Khanna, and Taylo 1998; Okonogi, Puttipipatkhachorn, and Yamamoto 2001; Mackenzie 1970; Kasraian et al. 1998); X-ray diffraction; or near infrared (Seyer and Luner 2001; Luner et al. 2000). Amorphous regions are thermodynamically unstable and exist in a higher energy state than crystalline regions within the particle (Darcy and Buckton 1998; Buckton and Darcy 1999). X-ray diffraction and DSC may detect large amounts of amorphous material. Smaller amounts of amorphous phase may be detected by moisture vapor sorption, isothermal microcalorimetry, and inverse gas chromatography (Buckton and Darcy 1999).

Mitscherlich's law of isomorphism describes in general terms the manner in which valency dictates isomorph formation according to the following expression for alums:

$$M'_2SO_4 \cdot M'''_2(SO_4)_3 \cdot 24\,H_2O \tag{2.4}$$

where M' is univalent (e.g., potassium or ammonium) and M''' is tervalent (e.g., aluminum, chromium, or iron).

Many phosphates, aresenates, sulfates, and selenates are isomorphic. Isomorphous materials can often form mixed crystals by co-crystallization from solutions.

## Enantiomorphs and Racemates

Enantiomorphs and racemates may be defined as two crystals of the same substance that are the mirror images of each other.

These crystals have neither planes of symmetry nor a center of symmetry.

Enantiomorphous crystals exhibit optical activity and are capable of rotating the plane of polarization of plane-polarized light. One form rotates to the left (laevo-rotatory, or L-form) and one to the right (dextro-rotatory, or D-form).

Molecules and substances that exhibit optical activity are described as chiral. Pasteur first isolated optical isomers of tartaric acid as shown in Figure 2.3. Mixtures of D and L crystals in solution will be optically inactive but can be resolved, unlike meso-form.

Crystalline racemates are normally considered to belong to one of two basic classes:

- Conglomerate: an equimolal mechanical mixture of two pure enantiomorphs;
- Racemic compound: an equimolal mixture of two enantiomers homogeneously distributed throughout the crystal lattice.

Racemates can be resolved manually or by forming a salt or an ester with an optically active base (usually an alkaloid) or alcohol followed by fractional distillation.

**Figure 2.3** Isomers of tartaric acid.

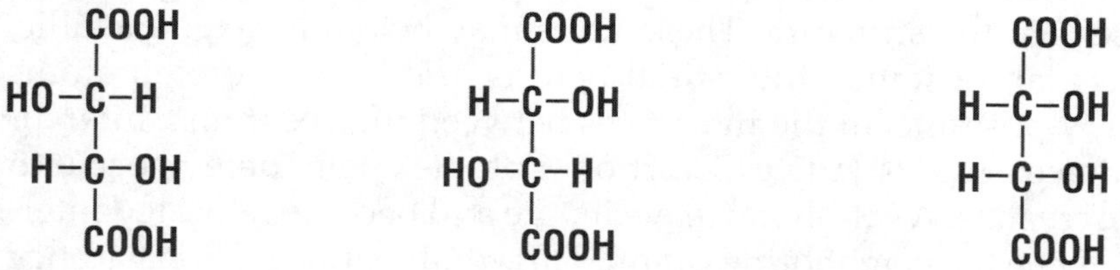


(a) Optically active tartaric acid

(b) Mesotartaric acid, not optically active

## Crystal Habit

Nearly all manufactured and natural crystals are distorted to some degree, and this may lead to misunderstanding of the term "symmetry." If crystal habit is considered as the product of restrictions on growth of crystals in a particular direction, then some generalizations can be made. No restrictions result in a crystal simply being a larger version of the unit cell lattice structure. For example, a cubic lattice will give rise to a large cubic crystal. Restrictions in one dimension will result in plate-like, tabular, or flaky particles. Restrictions in two dimensions yield needle-shaped acicular particles.

Rapid crystallization from super-cooled melts, supersaturated solutions, and vapors frequently produce tree-like formations called dendrites. Most metals crystallize in this manner, but due to the filling-in process, the final crystalline mass may show little outward appearance of dendrite formation.

## Crystal Imperfections

Most crystals are imperfect. Lattice imperfections and other defects can confer some important chemical and mechanical properties on crystalline materials. The three main types of lattice imperfection are point (zero-dimensional), line (one-dimensional), and surface (two-dimensional).

Vacancies are lattice sites from which units are missing, leaving "holes" in the structure. These units may be atoms (e.g., metallic, molecular, or ionic). Interstitial defects arise when foreign atoms occupy positions in the interstices between matrix atoms. Interstitial defects lead to lattice distortion. Complex point defects occur in ionic crystals. A cation can leave its site and become relocated interstitially near a neighboring cation. This combination (cation vacancy and interstitial cation) is called a Frenkel imperfection. A Schottky imperfection is a cation and anion vacancy. A substitutional impurity is a foreign atom that occupies the site of a matrix atom.

Two main types of line defect play a role in crystal growth: the edge and screw dislocations. Both of these are responsible for slip or shearing in crystals. Large numbers of dislocations occur in most crystals. Surface defects or imperfections involve the mismatch of boundaries and can be produced in crystalline materials as a result of mechanical or thermal stresses or irregular growth.

## Methods of Particle Production

Pharmaceutical particulates may be produced by constructive and destructive methods. Particle constructive methods include crystallization (Tiwary 2001; Umprayn et al. 2001; Tadayyan and Rohani 2000; Pham et al. 2001); spray-drying (Pham et al. 2001; Takeuchi et al. 2000; Sacchetti and Van Oort 1996; Kumar, Kang, and Yang 2001; Esposito et al. 2000; Dunbar, Concessio, and Hickey 1998); freeze-drying (Vavia and Adhage 2000; Rey and May 1999; Nuijen et al. 2001; Kirsch et al. 2001; Jameel, Amberry, and Pikal 2001; Jaeghere et al. 2000); and supercritical fluid techniques (Sacchetti and Van Oort 1996; Rogers, Johnston, and Williams 2001). Particle destructive methods include milling and grinding (Mura, Faucci, and Parrini 2001).

Crystallization is the most common method of particle production. The solid crystalline state is produced from a liquid solution by cooling, evaporation, precipitation, or the addition of another compound (solvent or solute) (David and Giron 1994). The process of crystallization involves nucleation followed by crystal growth. The crystalline structure of particles is affected by solvent, solute concentration, cooling rate, stirring, and the presence of additives.

Crystallization from solution essentially involves reaching a critical concentration in the solution, which represents significant supersaturation for spontaneous nucleation to occur in a homogeneous solution. In the presence of foreign particles (primary heterogenous nucleation) or nucleating crystals (secondary nucleation), marginal supersaturation is required. Figure 2.4 charts these nucleation situations.

Spray-drying is a process by which a dried particulate form is produced by spraying a liquid into a hot drying medium (Sacchetti and Van Oort 1996). It involves atomization of the liquid feed into fine droplets, evaporation of the solvent, and size separation of dried particles from the drying medium. Factors affecting the particle size, shape, and density of the final product include the feed concentration, solvent type, feed rate, atomization method, drying temperature, and residence time in the drying chamber.

**Figure 2.4** Crystallization from solution.

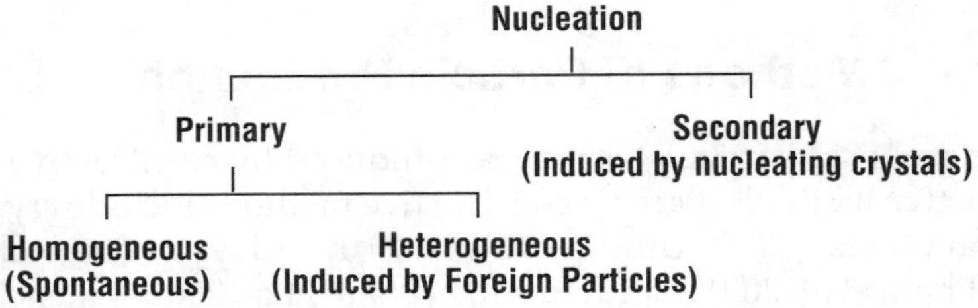


Freeze-drying, or lyophilization, is a process that involves freezing, sublimation, and secondary drying. Sublimation occurs when the water vapor partial pressure remains below the vapor pressure of ice. The solution composition, freezing temperature, and rate of freezing affect the sites and number of ice spaces. Sublimation is generally obtained by increased temperature at low pressure. The sublimation temperature should be optimized for the rate of sublimation, without the risk of melting. The sublimation pressure is adjusted to promote convection, generally 0.3 times the vapor pressure of ice at the temperature used (Bentejac 1994). The pressure is reduced in the secondary drying phase to eliminate adsorbed water.

Particle generation using supercritical fluids provides recrystallization by precipitation (Sacchetti and Van Oort 1996). Supercritical fluids exist at a region of high temperature and pressure at which the gas and liquid phase have the same density and appear as a single phase. A commonly used supercritical fluid is carbon dioxide, which has a low critical temperature (31°C), and is nontoxic, nonflammable, and inexpensive. Particle generation by supercritical fluid includes rapid expansion of supercritical solutions (RESS) and supercritical antisolvent (SAS) techniques.

Particle size reduction is generally performed by milling. The rate of size reduction depends on the material properties of the particle, starting size, the orientation in the mill, and the mill residence time. Compression, shear, and tension forces are used for size reduction (Lantz 1981). Compression forces produce crushing, shear forces produce cutting, and tension forces cause elongation and pull apart particles. Particles contain flaws, discontinuities, or imperfections in the structure. Application of milling forces at the flaws causes cracks and cleavage of the particles. Various types of milling equipment exist for size reduction of particles (Lantz 1981), including the fluid energy mill, ball mill, and hammer mill. The final particle size depends upon the energy input of the milling process. Fluid energy mills produce 5–30 μm particles, ball mills produce 10–200 μm particles, and hammer mills produce 50–400 μm particles.

## Particulate Systems

Particulate systems are disperse systems containing a solid phase with gas, liquid, or another solid (Table 2.3). Aerosols are solid particles or liquid droplets suspended in a gas. A complex aerosol may exist when solid particles are contained within the liquid droplets. A powder is a collection of small discrete solid particles in close contact with each other, with the void space usually filled with gas (Rietema 1991). Solid particles dispersed in a liquid phase are known as colloidal or coarse suspensions, according to the size of the dispersed particles. An emulsion contains liquid droplets dispersed in an immiscible liquid phase.

To understand the behavior of powders, the powder system needs to be conceived as a two-phase system of solid particles and a gas. The interparticle forces and properties of the gas phase play important roles in powder dynamics (Rietema 1991).

Colloidal systems (also known as sols) contain particle sizes of 1 nm to 0.5 μm (Martin 1993a). The three types of colloidal systems, classified according to the interaction of dispersed particles with the dispersion medium (Martin 1993a), are lyophilic, lyophobic, and association colloids. Lyophilic ("solvent-loving") colloids disperse

**Table 2.3** Types of Disperse Particulate Systems

| Disperse Phase | Continuous Phase | | |
|---|---|---|---|
| | Gas | Liquid | Solid |
| Gas | — | Foam | Solid foam |
| Liquid | Liquid aerosol | Emulsion | Solid emulsion |
| Solid | Solid aerosol | Suspension | Solid suspension |

easily with the dispersion medium. The attraction between the dispersed and dispersion phases leads to solvation or hydration in aqueous media. Lyophobic ("solvent-hating") colloids have little attraction to the dispersion medium. These colloidal systems are produced by using high-energy dispersion or condensation methods. Association colloids contain surfactants arranged as micelles. Spherical micelles are produced at surfactant concentrations close to the critical micelle concentration (cmc). Laminar micelles occur at higher concentrations. Although micelles are not solid particles, their size allows them to be considered as colloidal dispersions. Both lyophilic and association colloidal systems are thermodynamically stable, whereas lyophobic colloidal systems are thermodynamically unstable.

The presence and magnitude of electrical charge on colloidal particles affect the stability of colloidal systems. Like charges on colloid particles produce repulsion and prevent agglomeration, thereby increasing stability of the colloidal system. Oppositely charged colloid particles may aggregate when mixed.

Coarse suspensions contain particles larger than 0.5 μm (Martin 1993b). Pharmaceutical suspensions are usually coarse dispersions, in which insoluble particles are dispersed in a liquid medium. The desirable properties of an acceptable suspension include a slow settling rate of dispersed particles, ease of redispersion into a uniform mixture by shaking following settling, and sufficient viscosity for its application. These qualities are achieved by flocculated particles. Flocculated particles are weakly bonded, settle rapidly, do not form a cake, and are easily resuspended. Deflocculated particles settle slowly, but form a hard cake that is difficult to resuspend (Martin 1993b).

## Thermal Analyses

Thermogravimetric analysis and differential scanning calorimetry are the major forms of thermal analysis (Turi, Khanna, and Taylo 1988; Mackenzie 1970). Thermogravimetric analysis involves controlled heating of a sample to elevate its temperature while monitoring its mass using a microbalance. The weight loss is then recorded as a function of temperature. This is particularly useful for monitoring moisture loss and release of other volatile materials in addition to identifying the temperature for degradation of the substance being investigated. Differential scanning calorimetry involves direct comparison of the temperature of a test substance in an aluminum pan to that of a control of simply the pan over a range in which the pans are heated. This technique is useful in identifying phase transition temperatures and the magnitude of energy absorbed to facilitate the transition. As the test substance approaches a phase transition, applied heat is absorbed (endothermic response) by the solid material and used to mobilize atoms or molecules and rearrange them into new crystalline forms (polymorphs) or a new state of matter (liquid). This absorption results in no increase in temperature of the test substance apparatus with applied heat, whereas the control pan continues to rise in temperature as the heat is applied. Therefore, the difference between the two can be recorded as a developing peak whose area is proportional to the energy absorbed at a transition temperature. It is also possible for heat to be released (exothermic response) upon increasing temperature, which would also be detected by difference.

## Particle Surfaces

A number of techniques are available for probing the composition of the surfaces of materials of pharmaceutical interest. Some of the techniques, such as Auger electron spectroscopy (AES), secondary ion mass spectroscopy (SIMS), and X-ray photoelectron spectroscopy (XPS), can be combined with scanning electron microscopy (SEM) to provide chemical or elemental surface information to complement the direct imaging of the SEM. Another surface technique, gas adsorption, provides surface area information for particles.

### *Auger Electron Spectroscopy*

The Auger effect occurs due to nonradiative transitions in the inner electron shells of an atom (Ferguson 1989). When an energetic particle or photon strikes an atom, an inner shell electron may be ejected. Because this state is unstable, a higher-shell electron will fill the core hole vacancy. The energy difference from this transition will result in either the emission of a photon (usually an X-ray) or the ejection of an electron, called an Auger electron after Pierre Auger, who studied the effect. The radiative or nonradiative emissions each have a probability of occurring, and the sum of these probabilities must equal one. For exciting electron energies below 10 keV and for atoms with atomic numbers less than 33, the Auger electron emission is favored. The Auger electrons have energies unique to each atom. Therefore, by scanning the energy spectrum, the Auger electron peaks allow identification of all elements present except hydrogen and helium (Briggs and Seah 1990).

AES has a resolution down to 5 nm. Chemical information can also be obtained by AES but at a reduced spatial resolution. Auger electrons are low-energy particles, so they do not escape from deep in a molecule. Therefore, the sampling depth is limited to about three monolayers. Because an electron beam can be used to create Auger electrons, an Auger microprobe can be constructed by combining SEM with Auger electron detection. This provides a very convenient method to combine spatial and chemical information about particles.

### *X-Ray Photoelectron Spectroscopy*

XPS is related to AES, in which the energy of ejected electrons is determined by the measuring instrument. For surfaces irradiated with X-rays, the interaction of a photon with an electron in the atom can lead to the transfer of the photon energy to the electron (Riviere 1990). The photon energy, $h\nu$, must be greater than the electron binding energy, $E_B$, for the two to interact. Thus, the electron will be ejected with a kinetic energy $E_k = h\nu - E_B$. As for AES, the binding energy is unique to the atomic element, so elemental composition is provided.

Many commercial SEMs combine microscopy with XPS to probe the chemical composition of materials being imaged since SEMs measure emitted electrons. XPS provides information on the chemical bonding of surface materials more efficiently than does AES (Riviere 1990). Spatial resolution of XPS is on the order of 5 μm. All elements except hydrogen and helium can be detected. Like AES, XPS is limited to probing the first three monolayers of material. Fine-structure techniques, such as X-ray absorption fine structure (XAFS) and near-edge XAFS can probe the bulk of materials. However, these techniques require high-brightness X-ray sources, such as synchrotrons, and are, therefore, not very practical for pharmaceuticals.

### Secondary Ion Mass Spectroscopy

SIMS involves bombarding a surface with energetic primary ions (0.5–20 keV). Collisions of these ions with atoms in the sample result in the emission of ionized secondary particles that can be analyzed with a mass spectrometer (Wilson, Stevie, and Magee 1989). The data can be divided to provide mass and depth information. Unlike XPS and AES, SIMS can obtain bulk information, which provides the ability to detect impurities in the bulk. Spatial resolution is very good, down to 20 nm (Briggs and Seah 1990) since an ion beam can be focused by using a magnetic field. Commercial instruments are available that scan the ion beam over the sample, much like the SEM, to form an image of mass information.

### Gas Adsorption

Gas adsorption studies can be used to determine the surface area of an adsorbing solid (Grant 1993). Recent interest in porous particles for aerosol delivery makes this an especially important technique (Edwards et al. 1997). The nature of solid surfaces can be determined by observing the amount of gas adsorbed on a surface as a function of gas pressure. Plots of adsorption versus pressure are called adsorption isotherms. The amount of gas adsorbed in a surface reaction is often proportional to the surface area of the solid (Somorjai 1972). At low pressure or at the beginning of adsorption, the isotherm is linear. Clearly, however, the number of adsorption sites on a surface is limited. Brunauer, Emmett, and Teller (1938) proposed a set of isotherms based on the simple Langmuir isotherm to allow for adsorption in multilayers (the BET technique). By using different gases, the sizes of surface sites can thus be probed. The BET technique has been used to measure surfaces and describe them in terms of a fractal dimension in which the gas acts as a probe that varies in size (Avnir and Jaroniec 1989).

## Conclusion

The structure of crystalline materials has been characterized and can be defined in terms of 230 space lattices that fall into 32 classes of symmetry and seven systems that, in addition to the amorphous state, describe the range of particle morphology. Since regular geometries are a major simplification of true particle morphologies, the effects of specific defects that modify crystal structures to give rise to irregular forms are acknowledged.

Particles are produced by a variety of methods. Crystallization occurs from saturated solutions or hot melts to form solids. A variety of drying methods may be used to evaporate solvent and induce supersaturation. Crystals can be prepared by homogeneous or heterogeneous nucleation from solution. In the latter case, seed nuclei may be employed to control the form of the final product. Particles can also be modified by milling to reduce particle size by using high-energy input processes. Once particles have been prepared, they can be characterized by surface analytical methods to classify their structures and define them prior to their inclusion in pharmaceutical products.

The following chapters present the particle population characteristics as a means of predicting performance. These criteria can then be linked to the quality, efficacy, and safety of the product.

Particle morphology is a fundamental element in suspension and powder behavior. An understanding of the principles of particle formation and manufacturing process controls is essential in the optimization of product performance.

# 3

# Sampling and Measurement Biases

Sampling is a process whereby a small part of any population is selected for observation to estimate a given characteristic of the entire population. Ideally, sampling should be implemented such that although there is a reduction in the bulk, the population characteristics are unaffected (Gy 1982a). Although the process of sampling is of great importance, it is often neglected or undervalued. Sampling provides an opportunity to deliberately select a portion of a population while minimizing the influence of system variables (Thompson 1992a). Since estimates are inferred about the population characteristics from observations of a relatively small sample, the sample must be representative of the population (Bolton 1997b). The accuracy of population estimates depends critically on the characteristics of the sample being identical, within experimental limits, to those of the entire population (Dallavalle 1948f).

Each population contains a finite number of identifiable units. Each individual unit is associated with a fixed, invariant value of a certain characteristic. Thus, the population characteristic, known as the population parameter, is also a fixed value. Population parameters include the population mean ($\mu$), standard deviation ($\sigma$), and variance ($\sigma^2$). The population parameters would be known if the entire population was sampled and the methods of measurement were not subject to error or bias.

Sampling is used to estimate a population parameter. Sample statistics are quantities derived from the sample. Sample statistics include the sample mean ($\bar{x}$), standard deviation (s), variance ($s^2$), standard error of the mean (SEM), and coefficient of variance (CV), each of which approximates a population parameter. The statistics obtained from the sample are variable and depend on the particular sample chosen as well as the variability of the measurement (Thompson 1992a; Bolton 1997b). The uncertainty in estimates obtained from sampling arises from the observation of part of the population using methods with defined limits of accuracy and precision.

Sampling of particulate materials is performed for chemical and physical analysis in many stages of pharmaceutical manufacturing processes. These analyses ensure quality control of the final products. For example, raw materials may be examined for identification, potency, presence of impurities, and particle size distribution. In-process materials may be sampled for content uniformity. Final products are sampled for release tests, such as content uniformity, drug release, and dissolution. Unlike the sampling of miscible liquids, which form homogeneous mixtures at the molecular level, the sampling of particulate materials is difficult due to a variety of factors. Among these factors, particulate systems are composed of large units with respect to the population, unlike molecular dispersions. Dry powders are subject to aggregation and segregation, which may result in an inhomogeneous state. Liquid suspensions and particulate aerosols are subject to sedimentation and settling effects. Thus, sampling of particulate systems requires careful consideration to obtain representative samples.

## Sampling Strategies

A sampling strategy is the procedure by which the sample is derived from the population. A desirable sampling strategy provides an estimate that is close to the true value of the population characteristic. Careful attention to the sampling design enables unbiased estimates without assumptions about the population itself. An unbiased estimate is one in which the expected value over all possible samples that could be selected is equal to the true population value (Thompson 1992a). Accuracy is the closeness of the measured value or observation to the true value. An accurate measurement may be thought of as "unbiased," whereas a biased estimate is systematically higher or lower than the true value (Bolton 1997b). Samples that are not representative of the population may produce biased estimates (Thompson 1992a). Random error that shows no systematic bias but deviates from the population parameter is characterized by poor precision, where precision is defined as the variability of a set of measurements observed under similar experimental conditions. Sampling units may comprise a certain mass or volume of particles or individual dosage forms, such as tablets. The size of the unit and number of units sampled may affect the accuracy and precision of the population estimations (Thompson 1992a).

The sampling strategy should be chosen depending upon the experimental situation. Factors to be considered when choosing a sampling strategy include the nature of the population, the cost of sampling, convenience, and acceptable precision objectives related to the purpose of sampling. Sampling schemes may be divided into probability and nonprobability techniques (Bolton 1997c). Probability sampling gives all individual units in the population an equal probability of being selected, so the probability of being selected is uniformly distributed (Gy 1982i). Probability sampling methods include random, systematic, stratified, and cluster sampling (Figure 3.1).

Nonprobability sampling arises from convenience and simplicity rather than principles of statistical probability. Nonprobability samples are chosen in a particular manner and often are due to difficulty in controlling particular variables or lack of alternatives. For example, tablets may be sampled from only the top of a large container. Nonprobability sample methods often have hidden biases due to uncontrolled or unknown variables and are not recommended. Therefore, they are not discussed in this text.

## Random Sampling

Random sampling, as the name suggests involves the selection of individual samples at random, without any particular aim, pattern, or principle. Random sampling may be performed by assigning a

**Figure 3.1** Illustration of sampling schemes.

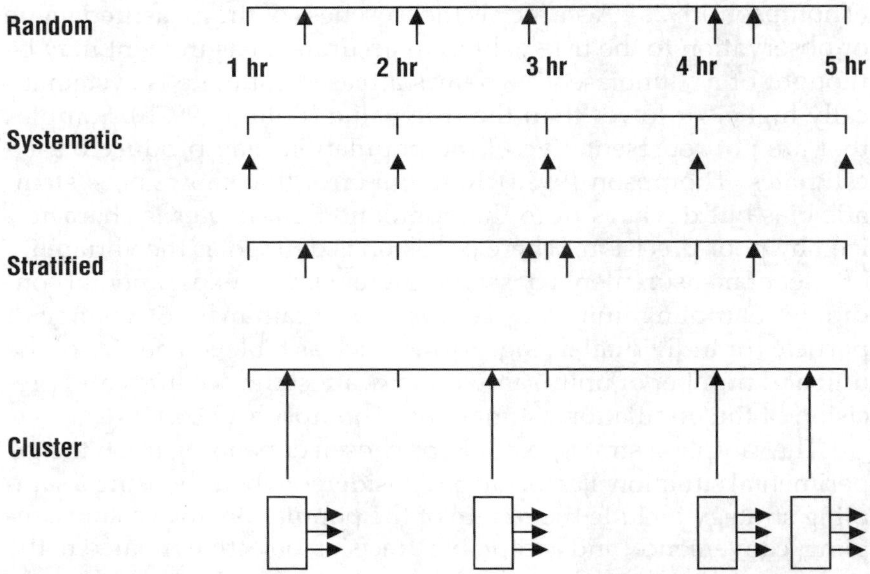


unique identification number to individual objects in the population and choosing the sample by randomly picking numbers or by using a table of random numbers (Bolton 1997c). Random sampling is a common method of selecting samples. Random sampling is less reproducible than the other schemes and may produce an undesirable allocation.

## Systematic Sampling

Systematic sampling involves the collection of individual samples at specified intervals, with the initial sample being selected in a random manner (Bolton 1997c). An example of systematic sampling is the collection of tablets at a particular time each hour during a production run. Systematic sampling is the most commonly used sampling scheme. It is very reproducible, easy, and convenient to implement. However, the use of a systematic sampling scheme in a population with cyclic or periodic behavior provides a potential risk of significant error and biased results. Periodic functions exhibit a regular pattern of alternating minima and maxima (Gy 1982f). Examples of possible periodic functions include cyclic patterns of human activity, automated processing equipment, and regulatory mechanisms. Although the sampling time interval is most unlikely to be an exact multiple of the periodic time interval associated with a particular bias, a stratified sampling scheme is recommended in the presence of periodic functions. Generally, systematic and stratified sampling methods provide the most reproducible selection.

## Stratified Sampling

In stratified sampling, the population is divided into subsets, or strata, and samples are selected at random from within each stratum. The random sampling of 10 tablets during each hour from a production run is an example of stratified sampling. Stratified sampling is easily implemented and reproducible, and it suppresses the inherent error caused by periodic functions (Gy 1982h). However, data analysis with this method is complicated because it should account for stratification. Stratified sampling is recommended when the strata are very different from each other (i.e., there is large between-strata variability), but the individual units within each stratum are similar (i.e., there is small within-strata variability) (Bolton 1997c).

## Cluster Sampling

Cluster sampling, also known as multistage sampling, is employed when many individual units are grouped together in larger clusters that may be subsampled. Clusters may be distinguished from strata in that they are discrete entities that may be sampled in their entirety. In single-stage cluster sampling, clusters are selected at random and all objects within the cluster are included in the sample. In two-stage cluster sampling, subsample units are chosen from the primary cluster (Bolton 1997c). For example, 100 bottles are randomly selected from a production batch of 10,000 bottles, and 10 tablets are randomly selected from 100 tablets contained in each bottle. In this example, the bottle is known as the primary unit, the tablets are the subsample units, and the population is the production batch of tablets.

## Statistical Analyses

Statistical analyses of a measured characteristic determined from samples allow the determination of the degree of accuracy of that characteristic for the bulk population (Dallavalle 1948f). Statistical analyses enable the differentiation between significant and nonsignificant differences by using the relationship between the probability of occurrence and the deviation from the established mean (Herdan et al. 1960d). Although any observation can never be absolutely excluded as a chance deviation from the mean value, a dividing line is drawn to distinguish "probable" from "improbable" deviations. Probability may be defined as the likelihood of an occurrence, expressed as the ratio of actual occurrences to that of all possible occurrences (Bolton 1997d). An arbitrary delimitation of a 5% probability level is commonly used to represent the borderline between such observations. In normal distributions, the 5% probability level represents a range of ±1.96 standard deviations from the mean. Any observed difference outside this range is considered as significant. The central limit theorem (CLT) states that the distribution of sample means taken from any distribution with a finite population variance and mean tends to be normal. The normal distribution is fully characterized by its mean ($\mu$) and standard deviation ($\sigma$), and all the properties of the distribution are known from these two parameters. The normal distribution is the basis of all modern statistical theory and methodology.

The confidence interval is a range of values in which the true population mean may be accurately approximated for a given probability. Since confidence intervals are calculated by using the sample mean and standard deviation, the confidence interval may change with different samples. Note that the calculated confidence interval for a given sample may not contain the true population mean (Bolton 1997e). Statistical hypothesis testing may be used to compare the sample mean to a known mean, sample means from two independent or related samples, or two sample variances. The t-test is used to determine any differences between two sample means. The chi-square ($\chi^2$) test is used to determine any differences between two sample variances. An analysis of variance (ANOVA) determines any differences among many groups by examining the between-treatment and within-treatment variances (Bolton 1997a).

## Sampling Techniques

Sampling techniques differ as functions of the size of the population being sampled. Incremental sampling techniques are implemented on unmovable lots, and sample-splitting methods are implemented on movable lots. A movable lot of particulate material is defined as one small enough to be handled in totality for the sole purpose of its sampling (Gy 1982e). Sample splitting, or dividing, reduces the sample size from gross samples (e.g., kilograms) to laboratory- or measurement-sized samples (e.g., grams) (Allen 1990c).

### Incremental Sampling

The incremental sampling process utilizes extraction of samples or increments from the lot by means of an extraction device. An example of incremental sampling is the sampling of a flowing stream at the discharge of a belt conveyor by using a cross-stream cutter. The four steps in the incremental sampling process are point selection, increment delimitation, increment extraction, and increment reunion (Gy 1982e). Point selection involves the selection of punctual increments at which sampling is to occur. The punctual increments are determined by the sampling strategy. During increment delimitation, the sampling device delimits or isolates the geometric boundaries of the sample, known as the extended increment. The extended increment thus surrounds the punctual increment. During increment extraction, the sampling device removes the particulate material contained within the extended increments. The particles or aggregates are extracted if their center of gravity falls within the boundaries of the extended increments. During increment reunion, the extracted particles are combined as an individual sample.

The sampling procedure must be designed to minimize the effects of segregation. Therefore, two rules should be followed during the sampling of bulk powders:

1. The powder should be sampled in motion.
2. The whole of the stream of powder should be taken for short time increments rather than taking part of the steam for the whole time (Allen 1990).

Collection of the whole moving stream for a short time offsets the effects of segregation. Biases are introduced when part of the stream is sampled instead of the whole of the stream (Gy 1982e). Sample bias or variations may be due to segregation, sampling equipment, and operator bias (Allen 1990c). Mechanisms of segregation will be discussed later in this chapter.

### *Sampling from Conveyor Belts*

The movement of powder on a conveyor belt is likely to produce two forms of segregation (Allen 1990c). When bulk powder is poured from a centrally placed feeder onto a conveyor, fine particles are concentrated in the center of the belt and coarse particles roll to the outer edges. In addition, vibration causes larger particles to rise to the top of the powder bed. Samples may be collected from the conveyor belt itself or as the material falls in a stream at the end of a conveyor belt. The stream should be sampled at the point of vertical trajectory (Gy 1982c). When samples are collected from either a stationary or moving conveyor belt, a section from the whole width of the belt must be removed. A thin layer of fine particles must not be left on the belt (Allen 1990c). Sampling from a stopped belt has been used as a reference method to check for sampling bias from a falling stream (Gy 1982b). When sampling from the stream falling from a conveyor belt, the whole stream should be collected where possible. Alternatively, the sample collector (cutter) should traverse the width of the stream.

The geometry and size of the cutter, cutter speed, and general layout also affect the sample collected (Gy 1982b). The two types of cutter geometry include straight and circular. Straight cutters have straight parallel edges that are perpendicular to the powder stream direction. Circular cutters have radial edges that converge toward the rotation axis and should be superimposable by rotation. Incorrect geometry includes rectangular, triangular, or trapezoidal openings. Deviation from correct edge orientation may result in delimitation errors (Gy 1982b). Bowing of the cutter edges may occur by the repeated hammering of coarse materials at high flow rates. This is prevented by mechanical adaptation to prevent damage and routine inspection to check for deformation. The accumulation of material on the inside of the cutter edges leads to nonparallel edges and eventual obstruction. This is prevented by routine inspection and cleaning. The capacity of bucket-type cutters should be at least three times as large as the volume of the intended increment to prevent overflow and bouncing. A minimum ratio of 20:1 for the width of the collector to the largest particle diameter prevents loss of large particles by bounce from the edges of the container (Allen 1990c). The cutter should provide a collection area at least three times the length of the stream thickness or diameter. The cutter speed across the stream should be uniform from one increment to the next. Particles may incorrectly bounce into the sample reject if the cutter is too narrow or moves too fast. Efficient emptying prevents particles remaining in the bucket after discharge. The general layout should allow sampling of the whole stream and prevent sampling occurring in the idle (nonsampling) position.

### *Sampling from Other Areas*

The entire contents of the bucket must be collected from a bucket conveyor. When sampling bulk material stored in bags, representative samples must be collected from several bags selected at random. Sampling from wagons and containers should be avoided, where possible, due to the severe segregation that inevitably occurs during filling and subsequent motion. If sampling from wagons and containers must occur, samples should be collected from at least 30 cm below the surface. Sliding of particles during their removal must be prevented (Allen 1990c).

## Sample Splitting

The four steps of the splitting process are fraction delimitation, fraction separation, fraction dealing out, and sample selection (Gy 1982j). During fraction delimitation, a splitting device isolates the geometrical boundaries of the population. During fraction separation, the material fractions are separated from the bulk population. This step is analogous to the extraction step in the increment process. During fraction dealing out, the material fractions are sectioned into a specified number of potential samples. Finally, during sample selection, all potential samples are submitted to the selection process to retain a certain number of samples. Methods of sample splitting include sample scoops or thieves, coning and quartering, table sampling, chute sampling, and spinning riffling.

### Sample Scoops

Sample scoops collect samples from the surface of bulk powder, whereas sample thieves (see next section) collect samples from the body of the material. Sample scoops and thieves are the most common methods for sampling stationary powder beds.

Fractional scooping is the simplest and cheapest bulk sampling method. It consists of moving the whole batch by means of a hand or mechanical scoop (Gy 1982j). The sample retained is one scoopful out of a total of N, enabling a splitting ratio of 1/N. Three types of fractional scooping include true fractional scooping, degenerate fractional scooping, and alternate scooping. In true fractional scooping, the scoopfuls are deposited into N distinct heaps. Following the complete dealing out of the whole batch, one or several samples are selected at random. In degenerate fractional scooping, the scoopfuls are dealt out, and every Nth scoopful is deposited in the sample heap. The remaining (N – 1) scoopfuls of the cycle are deposited in the reject heap. In alternate scooping, two heaps are formed and the sample heap is chosen at random following the completion of the scooping. This method is subject to operator bias, which may be suppressed by random selection of the sample fraction (Gy 1982j).

Sampling bias is common during the use of sample scoops as well as sampling thieves for a number of reasons. The primary reason is the violation of the two rules of sampling. Scoop sampling is particularly prone to error because the sample is taken from the surface, where segregation is likely to occur. The composition is likely to consist of larger particles and to differ from the bulk of the powder. As the scoop is removed, particles flow down the sloping surface of the powder. The percolation of fines through the coarse particles leads to a higher proportion of fines than coarse particles being withdrawn and larger particles being lost (Allen 1990c).

### *Sample Thieves*

Sample thieves are the only available method for sampling bulk blenders (Carstensen and Dali 1996). The two types of sample probes are side-sampling and end-sampling thieves. Side-sampling thieves contain a sample chamber within a tube. When rotated to the open position, the sample chamber aligns with the opening, thereby allowing the powder sample to flow into the cavity. Examples of side-sampling thieves include the grain thief, pocket thief, and slit thief. End-sampling thieves, also known as plug thieves, consist of a plunger inside a rod. The sample is obtained without the need for powder flow into the thief. Plug thieves cannot be used with poorly compressible powders because the sample would fall out of the tube as it is withdrawn from the powder bed. The diameter of the thief affects the tendency of samples to fall out (Garcia, Taylor, and Pande 1998).

Sampling bias of particulate systems by the use of sample thieves has been revealed by mixture solidification and image analysis techniques (Berman, Schoeneman, and Shelton 1996). Sampling bias is likely to occur when small samples are extracted with a thief from large volume populations for a variety of reasons, such as the following: Significant disturbances are produced during insertion of the sample thief into the static powder bed, which may result in interparticulate movement and segregation. The insertion of the thief into the powder bed may cause contamination of the lower regions with powder from the upper regions of the bed (Garcia, Taylor, and Pande 1998). Powder compaction during the insertion into the powder bed may inhibit powder flow into the sample chamber. Thief insertion may also cause particle attrition (Berman, Schoeneman, and Shelton 1996). The static pressure at the sampling location may affect powder flow into the sample chamber because greater pressure is observed at lower regions of the powder bed.

The flow of particles into the sample chamber of the sample thief introduces further uncertainty. Preferential segregation during thief sampling may arise due to particle size differences between drug and excipient particles as well as electrostatic properties. Larger particles (generally excipient) may preferentially flow into the cavity due to their enhanced flow properties (Garcia, Taylor, and Pande

1998). In addition, free-flowing particles have been observed to enter the sample cavity of side-sampling probes before the thief was opened. Particles may be entrapped and cause difficulty in sample chamber closing (Allen 1990c). Particles entrapped in the space surrounding a closed sample chamber are captured when the sample chamber is opened (Carstensen and Dali 1996).

The sampling technique may influence sampling bias when using sample thieves. Significant differences in the particle size distribution of sampled powders have been observed with sample depth, sample thief type, and sample chamber orientation. The angle of insertion may affect the flow dynamics of powder entering the sample chamber (Berman, Schoeneman, and Shelton 1996). Differences in powder sampling have been observed with different types of sample thieves. Tighter relative standard deviation (RSD) values have been obtained when using the plug thief compared with the use of a grain thief (Garcia, Taylor, and Pande 1998). However, difficulty in obtaining reproducible desired sample weights was observed for the plug thief.

Methods of minimizing sampling bias include improving thief design and using consistent sampling techniques. The sampling thief should be primed to minimize any preferential adherence of active or excipient material to the walls of the sample cavity. This is performed by inserting the thief in a blend location that is not being sampled, withdrawing several samples, and returning them to the blend (Mohan et al. 1997).

A core sampler, consisting of a large cylindrical tube with a shapely tapered end, has been developed as an improved thief for powder sampling (Muzzio et al. 1999). When the core sampler is inserted into the stationary powder bed, an undisturbed cylindrical core of powder is isolated within it. Cohesive powders may be extracted without draining due to their interparticulate forces. Free-flowing powders are removed with the aid of a capping fixture. The extracted core tube may be subdivided into a number of samples with a rotating screw discharge device. The insertion of the core sampler disrupts the powder on the outside of the sampler, so the minimum recommended separation between sampling locations is five core diameters with the capping device, or two core diameters without the capping device. In addition, removal of the core sampler further disrupts the exterior powder bed. When multiple locations are sampled, all samplers should be inserted before any of them are removed to ensure that disturbances during removal of the core tubes do not affect the isolated powder samples within each tube.

### *Coning and Quartering*

Coning and quartering, also known as "Cornish quartering," is the oldest probability sampling method, dating back to the early nineteenth century when it was used in the tin mines in Cornwall (Gy 1982j). Coning and quartering relies on the radial symmetry of the conical heap to produce four identical samples (Allen 1990c). The sample is poured into a conical heap from the central apex. This cone is flattened into a flat circular cake, retaining the symmetry achieved in the first step. The flattened powder bed is divided into four quarters by using a cross-shaped metal cutter. Two opposite quarters are retained, and the other two quarters are rejected. Reliable results are obtained if the heap is symmetrical about a vertical axis and the cutting planes align along the axis. However, in practice, the heap is unlikely to be truly symmetrical, resulting in different-sized samples (Allen 1990c). This method is very time consuming and subject to operator bias; therefore, it is not recommended. It is no more accurate or precise than alternate shoveling (Gy 1982j).

### *Table Sampling*

In table sampling, a series of holes and prisms is placed in the path of the stream to fractionate the powder. The powder is fed onto an inclined plane and is directed by strategically located prisms toward the holes or the sampling chute at the end of the plane. The sample is taken from the end of the plane, and the powder that falls through the holes is discarded. The accuracy of this method depends on the uniformity of the initial feed. Errors are further compounded by the sequential removal of parts of the stream, resulting in low accuracy (Allen 1990c).

### *Chute Splitter*

The chute splitter consists of an assembly comprising an even number of identical adjacent chutes, each of which leads the powder alternately toward two different collection buckets (Gy 1982j). The sample is poured into the top of the chute and repeatedly halved until the desired sample size is achieved. Chute splitters are of various types, including a V-shaped trough divider, oscillating hopper divider, and oscillating paddle divider (Allen 1990c). This method is inexpensive and easy to use, and it gives accurate sample division when used properly. However, it is subject to operator bias. Potential bias may be suppressed by alternating the bucket chosen for sample retention (Gy 1982j). Overflow of one of the chutes may occur if material is discharged too quickly or close to one of the chutes.

### *Spinning Riffler*

The spinning riffler fractionates the powder into revolving collection containers under a stationary feeder. The powder should be filled into a mass flow hopper (steep sides at an angle greater than 70°) with minimal segregation. This is achieved by moving the pour point to prevent heap formation. The powder flows from the hopper onto a vibratory feeder, and then finally into collection containers on a revolving table. The powder flow rate and collection stage revolutions should be maintained at a uniform speed to ensure similar fraction weights. Each container should receive more than 30 increments (Gy 1982j). Optimum results are obtained by using at least 35 collection containers (Allen 1990c). This method conforms to the two rules of sampling, and its use is recommended whenever possible (Allen 1990c). It was found to be superior when compared to other methods. However, this method can be used only for free-flowing powders.

## Segregation

Segregation is observed only in particles that have freedom of movement and differences in such physical properties as particle size and density (Chowhan 1995a). During segregation, particles with the same physical properties accumulate in one part of the mixture (Rhodes 1998). The segregation of powders may be predicted by their flow properties. Free-flowing powders have the highest tendency to segregate because the particles separate easily. Cohesive powders have a reduced tendency for segregation due to the strong interparticulate forces inhibiting motion of particles relative to one another.

The major factors affecting segregation include the particle size and its distribution, particle density and its distribution, and particle shape and its distribution. The minor factors affecting segregation include the coefficient of friction; surface roughness (rugosity); moisture content; shape composition and surface of the container; and differences in resilience of the particles (Chowhan 1995a).

Size segregation is the most prevalent type of segregation. Particle interference and the influence on powder packing are the proposed causes of size segregation (Chowhan 1995a). Particle interference occurs when the average unobstructed distance between particles of any given size is smaller than the diameter of the next smaller size. Particle interference affects the packing of the powder bed. Segregation occurs during stirring or vibration of the powder mixture when the void space between the larger particles is smaller than the space between smaller particles.

Segregation of particles due to density differences occurs during acceleration and deceleration of particle motion. The ratio of density difference in granular materials is generally smaller than 3 to 1, so the degree of segregation due to density is small compared with that due to size. Density segregation is also affected by particle size. Segregation was not observed in equal-sized spheres with a three-fold difference in density, but rapid segregation was produced with small dense particles, and retarded segregation occurred with large dense particles (Chowhan 1995a).

The influence of particle shape on segregation may be explained by differences in particle motion, such as rolling and sliding. Particles with rounded shapes and smooth surfaces slide away from rough particles on a smooth surface or vibrating feeder.

Segregation may occur during powder mixing, transport, storage, handling, and sampling processes. The mixing process has an opposing action to segregation and is also influenced by the flow characteristics of powders. Equilibrium may be attained at the point where the rate of mixing equals the opposing rate of segregation (Chowhan 1995a). When powders are mixed, the homogeneity initially improves rapidly, then deteriorates, and finally is followed by a period of cycling. A homogeneous powder blend obtained during mixing may segregate during subsequent handling, storage, and transport processes.

Segregation may occur within the powder bed, in a powder heap, and during fluidization (Chowhan 1995a). In a powder bed, when subjected to vibration or stirring, larger particles rise to the upper level of the powder bed and smaller particles sink to the bottom, regardless of density. Segregation occurs mostly in the top 15 cm of the bed. Maximum segregation occurs when the angle of vibration is around 30° to the horizontal. Layer loading or the use of compartments with a small cross-sectional area may minimize segregation within a powder bed (Chowhan 1995a).

The formation of a powder heap results in surface segregation. As the particles are poured onto the top of the heap, a very high velocity gradient forms in the first few layers of the surface. This layer effectively forms a screen that allows fine particles to percolate to the stationary region in the center of the cone while coarse particles roll down the outside of the heap. For example, when a hopper is filled, coarse material runs down the cone to the walls and the fine material forms a central core. On discharge of the powder, the fine particles are discharged first, followed by the coarse particles, which roll down the inverted cone that has formed. Heap formation is the most frequent cause of size segregation. This type of segregation is minimized by keeping the top surface of the powder bed level as it is filled or by reducing the cross-sectional area (Chowhan 1995a).

During fluidization, small particles are carried upward in the wake of gas bubbles. Larger and denser particles are insufficiently supported by the bubbles and fall through to the bottom. Small dense particles may descend by percolation, whereby they slip into the interstitial space between larger particles. Segregation during fluidization is restricted to regions disturbed by bubbles. The velocity of the fluidizing gas determines the degree of segregation (Chowhan 1995a). The gravitational settling of particles in laminar conditions may be described by using Stokes' law. The effect of particle size on fluidization is discussed in Chapter 5.

Segregation has implications in powder handling processes. Segregation mechanisms must be understood to enable the collection of representative samples. The manufacture of pharmaceutical particulate systems is affected greatly by segregation. Size segregation results in fluctuations in size distribution of powder mixtures and variations in bulk density. Tableting and encapsulating machines measure dosage size by powder volume and assume constant bulk density, so segregation is one of the major causes of weight variation in tablets or capsules (Chowhan 1995a). Segregation may also result in variations in chemical composition of the powder mixture, due to the behavior of particles of different substance, with implications for the content uniformity of final products (Rhodes 1998).

The four main mechanisms of segregation include trajectory, percolation, vibration, and elutriation segregation. Trajectory segregation occurs when particles with different physical properties are subjected to horizontal motion. When projected horizontally, larger particles travel farther before coming to rest due to differences in mass and inertia (Rhodes 1998). This mimics the principle of elutriation.

Percolation segregation occurs by gravitational force during such movement of the powder bed, as vibration, pouring, and shaking (Rhodes 1998). The smaller particles fall down through gaps during rearrangement of the powder bed. This may also occur if the powder bed has a high void fraction (low bulk density) (Chowhan 1995a).

Vibration segregation is increased at higher amplitudes and lower frequencies. The rate of vibration segregation increases with lighter particles (Chowhan 1995a). However, the mechanism for this effect is not entirely understood. Two possible explanations exist. First, the momentum of large particles during the upward motion causes penetration of the loosely packed fine particles above. During the downward motion, large particles cause compaction of the fine particles below, creating resistance to the downward motion of the large particle. The large particle thus essentially moves up "one notch" during each vibration cycle. The other possible mechanism is the action of the container walls producing a convection pattern with the fine particles rising in the center and falling near the walls. The convection motion of the fine particles lifts the larger particles to the top of the powder bed (Rhodes 1998). This is similar to the principle of boiling in a vertical tube.

Elutriation segregation occurs due to upward motion of a fluidizing gas. When a powder with a large proportion of particles smaller than 50 μm is poured into a storage vessel or hopper, air is displaced upward. The upward velocity of the air may exceed the terminal free-fall velocity of the finer particles, which may remain in suspension after the larger particles have settled; this results in a pocket of fine particles being generated each time the vessel is filled (Rhodes 1998).

Two other segregation mechanisms are limited to the segregation of interactive mixtures: interactive-unit segregation and constituent segregation. Interactive mixtures, also known as ordered mixtures, are powder mixtures in which the fine particles (usually drug particles) adhere to larger carrier particles (usually excipient particles). Interactive-unit segregation is caused by different-sized carrier particles. This results in different proportions of fine adherent particles on the surface of the carrier particles and leads to fractions containing higher or lower drug concentrations within the powder mixture (Chowhan 1995b). Constituent segregation occurs when adherent fine particles are separated from the coarse carrier particles or when carrier particles are unable to bind all the fine particle components present in the mixture. This results in free fine particles moving independently of the carrier, producing segregation by percolation or micro-fluidization (Chowhan 1995b).

## Sampling

Sampling of powders occurs in many stages in pharmaceutical development and production. The types of materials sampled range from raw materials to in-process powder blends to final products. Raw materials must be tested for identity, strength, purity, and quality prior to use for manufacturing. Unacceptable raw materials produce unacceptable final products. This section deals specifically with sampling of powder blends, liquid suspensions, particulate aerosols, and final products.

### Powder Blends

The primary aim of any blending process is to provide a uniform distribution of drug particles in a powder blend, as either a random or interactive mixture. The homogeneity of blends must be achieved and maintained to ensure that each dosage unit of a formulation contains the same quantity of medicament (Lantz and Schwartz 1981; Muzzio et al. 1997). This is especially important in formulations containing small amounts of high-potency components. The content uniformity of the final product is adversely affected if the blend homogeneity is below a critical point. Blend uniformity testing determines that the active ingredient is homogeneously dispersed throughout the powder blend. Due to the potential for segregation to occur during subsequent processing and storage, acceptance criteria ought to be tighter for powder blends than for finished products.

Legal precedent, current regulations, and guidelines provide recommendations on how blend uniformity testing should be carried out. Current good manufacturing practice (cGMP) regulation 21 CFR 211.110(a) states that the sampling of in-process materials and drug products is required, to ensure batch uniformity and integrity of drug products (CFR 2000). The monitoring of the adequacy of mixing is required to ensure uniformity and homogeneity. The Barr decision of February 1993 (*U.S. District Court for the District of New Jersey v. Barr Laboratories, Inc.,* Civil Action No. 92-1744) stated that the appropriate sample size for blend content uniformity in both validation and ordinary production is three times the dosage size of the active ingredient. Powder samples may be taken from either the blender or the transport drum, and the sampling locations must be chosen to provide a representative cross section of the blend. These locations should include areas that have the greatest potential to be nonuniform (Berman and Planchard 1995). The draft guidance for blend uniformity analysis of abbreviated new drug applications (FDA 1999) recommends routine in-process blend uniformity analysis for each production batch of drug products for which the United States Pharmacopeia (USP) requires content uniformity analysis. This includes products that contain less than 50 mg active ingredient per dosage form unit or when the active ingredient is less than 50 percent of the dosage form unit by weight. The recommended sample size of the powder blend is no more than three times the weight of an individual dose. Sample sizes of no more than 10 dosage units may be collected, if sampling bias is experienced. The recommended acceptance criteria that the mean content of active ingredient falls within 90.0 to 110.0 percent of the expected quantity and the RSD is no more than 5.0 percent. [An RSD of 5 percent provides a 95 percent chance of samples falling within 10 percent of the mean (Crooks and Ho 1976).] This allows for any potential loss in blend uniformity during subsequent manufacturing steps. At present, this draft has not been implemented.

In response to the guidance draft, many pharmaceutical companies have expressed concern over the value of blend uniformity testing as a routine in-process test for production batches. Although blend uniformity testing during development and validation processes ensures the consistency and accuracy of the mixing process, routine blend uniformity testing in commercial production is viewed as redundant because finished dosage units are tested for drug uniformity. Concerns also arise from the possibility that products will unnecessarily fail the stringent acceptance criterion due to sampling bias caused by the difficulties in obtaining small samples from a blender by using a sampling thief.

Unit-dose samples of one to three times the finished dosage weight sampled from the mixer or drum by using a sample thief may not be representative of the entire batch due to sampling bias (Mohan et al. 1997). Inconsistencies have been observed in the content uniformity of final blends compared to that of corresponding finished products. These inconsistencies have been largely attributed to sampling bias due to the use of thieves for unit-dose sampling. Generally, the drug content of powder blends sampled with a thief is lower than the content uniformity obtained from corresponding tablets (Carstensen and Dali 1996; Berman, Schoeneman, and Shelton 1996; Berman and Planchard 1995). Larger variability is observed in powder blends compared with the corresponding tablets (Garcia, Taylor, and Pande 1998; Mohan et al. 1997).

The Product Quality Research Institute (PQRI) blend uniformity working group has proposed the use of stratified sampling of blend and in-process dosage units to demonstrate adequacy of mix for powder blends specifically applicable for validation and production batches for solid oral drug products. This proposal arose from the limitations in current sampling technology and subsequent powder segregation during handling, which reduces the effectiveness of blend sample analysis to ensure adequacy of blending. The sampling and analysis of in-process dosage units is proposed as an alternative to routine blend sample analysis. Stratified sampling selects units from specifically targeted locations in the blender or in the compression and filling operation that have a higher risk of producing failing content uniformity results. The analysis of in-process dosage units provides an accurate measurement of homogeneity of the product. It also eliminates sampling errors related to thief sampling and weighing errors of blend samples. The detection of subsequent segregation following blending is improved. Variance component analysis of the data enables the determination of the variability attributed to the blend uniformity and any sampling error. High between-location variation indicates poor blend uniformity, whereas high within-location variation indicates sampling error. In addition, dosage unit analysis satisfies the cGMP 21 CFR 211.110 (a)(3) requirement (CFR 2000) by indirectly measuring the uniformity of the blend by sampling and testing in-process dosage units.

Blend uniformity testing may be a useful tool in the development and optimization of manufacturing processes. Blend uniformity testing in production processes aims to minimize the likelihood of releasing a batch that does not meet end-product specifications. However, accurate characterization of the powder mixture may not be feasible by using a sample thief due to sampling biases. Blend homogeneity may not be indicative of the content uniformity of the final product due to further mixing or segregation processes that may occur during subsequent handling. The analysis of blend uniformity at earlier stages serves only as a coarse indicator of potential problems (Muzzio et al. 1997).

During blending process development and validation, sampling of the blender and intermediate bulk containers is required for the optimization of blending time and speeds. Sampling enables the identification of dead spots in the blender, segregation in the container, and the presence of sampling error. The type of blenders commonly used are tumbling and convective blenders. Tumbling blenders rotate around a central shaft. As these blenders rotate, the powder tumbles downward following a cascading motion. The powder flow patterns within tumbling blenders are complicated with two distinct regions of tight-packing and expanded volume. Particles move in recirculating flow patterns. Long blending times are required for particles entrapped in "dead regions" to mix with the remainder of the system. Convective blenders consist of a stationary vessel with a rotating impeller (Muzzio et al. 1997).

Various mixing indices have been used to describe the homogeneity of powder blends, blending equipment, and procedures (Lantz and Schwartz 1981). They generally involve the comparison of the measured standard deviation of the samples with the estimated standard deviation of a completely random mixture. Common parameters used to describe the degree of homogeniety include the sample standard deviation, variance, and relative standard deviation. The degree of mixing is determined by comparing the mean value of the sample analysis with the target value of the mixture. The sample standard deviation gives an indication of the uniformity of the samples. Differences between the assayed mean content and theoretical mean content do not solely indicate a lack of homogeneity. Differences may also result from poor or inadequate sampling, improper handling of the powder sample, or errors in the assay method (Lantz and Schwartz 1981). In addition, errors resulting from the limits of analytical quantitation may further compound the sampling error, as discussed later in this chapter.

Various statistical methods for the evaluation of blend uniformity have been proposed. A method based on statistical tolerance limits provides a 95 percent confidence level that at least 90 percent of the values of the entire population are within the lower and upper limits constrained at 85 percent and 115 percent of blend uniformity (Guentensberger et al. 1996). Two other statistical methods include Bergum's method and the standard deviation prediction interval (Berman and Planchard 1995).

The sample size, number of samples, and location of sampling may affect measurement values. The sample size should approximate the unit dose size of the final product (Lantz and Schwartz 1981). The number of samples taken from a random mixture should be no less than 20. Larger sample numbers provide greater confidence in estimating the true mean and standard deviation of the mixture. The number of samples required for examination depends upon the uniformity of the mixture. One sample may be sufficient for the characterization of a homogeneous powder mixture, whereas multiple samples may be required for heterogeneous mixtures (Dallavalle 1948). The number of samples is the dominant factor in determining the accuracy of the characterization (Muzzio et al. 1997). The sample size is important only when it is of similar magnitude to the scale of segregation of the system. For poorly mixed systems, the sample size is irrelevant and has no effect on the characterization of the mixture (Muzzio et al. 1997). Sampling locations should be identified prior to the blending operation. At least 10 locations are required for tumbling blenders (i.e., V-blenders, double cone, drum mixers), and at least 20 locations are required for convective blenders (i.e., ribbon blenders). Potential areas of poor blending, such as the corners and discharge areas of ribbon blenders, should be selected. All samples should be removed from the blend at the same time. Resampling of the blend is not recommended.

## Liquid Suspensions

The nature of suspensions requires complete redispersion of suspended particles prior to sampling. The gravitational settling of suspended coarse particles during storage is governed by Stokes' law. For the sample to be representative of the whole population, the suspension should be sampled while uniformly dispersed and prior to particle settling. Liquid suspensions should be sampled by using a syringe or pipette. This method is useful for sampling of fine particles; however, large particles may sediment due to gravitational settling. Commercial suspension samplers are available for samples of 2 to 10 mL (Allen 1990c).

## Particulate Aerosols

The principles involved in the sampling of particulate aerosols arise mainly from air pollution control. As emphasized previously, it is essential to obtain samples that are representative of the whole population. Two main difficulties are involved in the removal of representative samples from gaseous suspensions of particles. First, aerosols are inherently unstable, and pharmaceutical aerosols rarely approach equilibrium. Second, any disturbance to the airflow affects particle motion within the aerosol. When the entire aerosol cannot be sampled, the aerosol should be sampled under isokinetic conditions, in which the air velocity at the point of sampling is identical to that of the aerosol. This ensures no disturbances of the gas streamlines (Allen 1990b). When the sampling velocity differs from that of the gas stream (anisokinetic sampling), particles are deflected from their original direction of motion, depending on their mass, due to the disturbance in the gas streamlines. The sampling velocity affects both the sample mass and size of particles in the sample collected.

Gas sampling equipment consists of a set of nozzles attached to a sampling tube and a collector. The nozzles remove particulate matter from gas samples. The sampling probe should be isoaxial, parallel to the flow lines, to prevent concentration loss of particles. Sampling nozzles should have circular cross sections. The thickness of the open edge should be less than 1.3 mm. The internal and external surfaces should have an inclination less than 45° to the axis of the nozzle, and the nozzle diameter chosen to maintain isokinetic conditions (Allen 1990b).

The conditions at the point of gas sampling, such as flow rate, pressure, temperature, and the volume of gas sampled, should be measured. The particle content of a gas may be expressed as the mass flow rate, which is the mass of particles passing across a given area per unit time, or the particle concentration, which is the mass of particles per unit volume of gas (Allen 1990b).

Aerosol particles may be collected by deposition in a cyclone, gravitational settling, filtration, impaction or impingement, and thermal or electrostatic precipitation. Gravitational settling involves the collection of particles from a known volume of aerosol onto collection plates enclosed in a chamber. Impaction involves the deposition of particles from a moving airstream onto a plane surface. Thermal precipitation employs repulsion of particles from a heated surface and collection on a cold surface. Charged particles are deposited on a collecting anode following the charging of particles by a corona effect from an ionizing cathode at high potential to achieve electrostatic precipitation.

Pharmaceutical aerosols are generally characterized using inertial impaction techniques, such as cascade impaction. In this technique, the whole aerosol is sampled, which eliminates the risk of the sampled portion not being representative of the bulk population. However, caution must be exercised to ensure that the characteristic properties are representative of the aerosol properties. For example, the airflow rate at which the aerosol is generated may not be identical to the specified airflow rate required for measurement.

## Final Products

Various release tests are required for final products, including disintegration, dissolution, drug release, content and weight uniformity, and stability tests. The sampling processes for each of these tests are generally governed by pharmacopeial monographs, such as British,

European, or U.S. Pharmacopeia; and regulating bodies, such as the U.S. Food and Drug Administration (FDA), the British Medicines Control Agency (MCA), or the European Agency for the Evaluation of Medicinal Products (EMEA). For example, the dissolution test for solid dosage forms (tablets or capsules) requires the testing of 6 units from a production batch, with a further 6 and 12 units tested if the initial test results fail the acceptance criteria, which vary slightly between the British and U.S. Pharmacopoeias.

## Sampling Errors

Errors in the estimation of population parameters may be sampling or nonsampling errors. Nonsampling errors include the errors that occur during sample preparation, analysis, or measurement. Sources of variation in experimental observation include the sample, instrumentation, operator, and background noise, which includes such environmental variables as temperature and humidity (Bolton 1997b). If it is assumed that sample preparation and measurement are performed without error, the difference between a sample characteristic and its corresponding population parameter would be due to sampling error (Gy 1982d).

Specific sampling errors introduced by the various sampling techniques have already been discussed. Generally, sampling errors in particulate systems arise from the heterogeneity of the population. The heterogeneity of a discrete population may be further classified into constitution and distribution heterogeneity (Gy 1982e). Constitution heterogeneity accounts for the intrinsic properties of the population, whereas distribution heterogeneity accounts for the properties of the spatial distribution of the particles throughout the lot. Thus, variations in sample measurement may be caused by the intrinsic variation within the population or variation due to sampling. Other sampling errors involve incorrect sample delimitation and extraction (Gy 1982e). For example, the metal cutter not being placed symmetrically on the vertical axis, thus resulting in unequal quarters of powder, illustrates incorrect delimitation during coning and quartering. Incorrect extraction occurs if the delimited sample is not fully removed or if the sample contains other particles that should not be in the sample. Another example is sample bias due to incorrect sample extraction if a sample thief is not completely emptied. This may also result in the contamination of the subsequent sample if the sample thief is not cleaned before sampling.

## Preparation Errors

The sampling methods described in the previous sections relate to the reduction of large bulk powders to a sample size that may be conveniently sent to the laboratory for analysis. This sample may be on the order of a few grams to a kilogram. Generally, further reduction of the mass of powder is required for the analysis or measurement. The mass of powder actually required depends upon the assay or characterization technique used. Microscopy requires only a few milligrams, whereas sieving requires 25 to 50 g. The technique commonly used to extract a measurement-sized sample from a laboratory sample is to scoop out an appropriate amount with a spatula. Sometimes the bottle is shaken or stirred in an attempt to mix the particles within it. However, this approach is subject to segregation and bias. Segregation is likely to occur during the shaking or stirring of the powder. The sample removed by the spatula will include mostly surface particles, which may be different in composition than the bulk powder. The removal of the spatula will result in size segregation, in which fine particles are captured and large particles are lost after flowing down the sloping surface. This effect is especially important if a flat spatula is used. Methods recommended for the reduction of laboratory-sized samples to measurement-sized samples include the spinning riffler, coning and quartering a paste, and specific shaking techniques (Allen 1990c).

The analysis of the samples may be performed on individual samples, an aliquot of each individual sample, pooled samples, or an aliquot of pooled samples. When samples are pooled, equal amounts of individual samples are required to prevent bias (Dallavalle 1948f). However, analysis of the whole sample is recommended whenever possible.

Other sources of error during sample preparation include contamination, loss of sample, and alteration of chemical or physical composition. Intentional or unintentional mistakes, such as operator error due to ignorance, carelessness, or lack of experience, may also occur (Gy 1982g). Errors also may arise from inappropriate storage of samples, such as near sources of low-frequency vibration (e.g., refrigerators or air conditioning units), under extremes of temperature and humidity, or next to doors that are opened and closed frequently.

The sample itself may be contaminated from a variety of sources, including the particulate system being sampled, external material, and abraded or corroded processing or sampling equipment (Gy 1982g). Contamination may occur from the dust of fine particles contained within the particulate systems. Reducing free falling of dry powders will also reduce dust formation. In addition, contamination from dust may be reduced by using dust collection systems and protection covers on the sampling devices. Cleaning the sampling circuit and sampling devices prior to new sampling operations prevents the contamination of any external material. Stray material from abraded or corroded equipment may cause contamination of the sample, which can be prevented by selection of appropriate materials, such as stainless steel for all machinery parts in contact with the material sampled. Glass shards from containers may be present in samples if poorly handled.

The loss of sample material may result in an underestimation of a population parameter, such as drug content. Examples of lost material include fine particles as dust, materials remaining in the sampling or preparation equipment, or certain fractions of the sample (Gy 1982g). Loss of sample may also result from incomplete dissolution of the sample prior to analysis. Dissolution of poorly soluble compounds may be aided by sonication. The presence of undissolved particles may be viewed using a polarizing light source.

Chemical processes, including oxidation or hydrolysis, may alter the chemical composition of a sample. The adsorption of water or other gases from the atmosphere may affect the chemical nature of a sample. Moisture may be removed by drying in an oven if the sample is not heat-sensitive. Moisture adsorption may be prevented by storage under vacuum or in a dessicator (Gy 1982g). The use of an appropriate dessicant is required. Silica gel is recommended for low to intermediate equilibrium moisture content materials, and phosphorus pentoxide is recommended for high moisture content materials. The physical composition of a sample may be altered during transport or processing. For example, the particle size may be altered due to comminution of particles during free-falling motion (Gy 1982g).

## Particle Size Measurement Errors

Measurement errors may occur in all types of characterization or analysis. Such errors may be due to intentional or nonintentional operator mistakes or instrument-specific measurement bias. Particle size measurement is subject to various types of error, including sampling and experimental preparation and measurement errors. Discrepancies in the measurement of particle size may arise due to differences in the sample examined and the physical dispersion of the sample (Herdan et al. 1960e). Differences in particle size distribution obtained from different techniques arise from differences in the degree of resolution and size range, the measurement principle used for size determination, the type of distribution (number or weight), and the measurement procedure (single or multiple particles).

The importance of the sample being representative of the particle size analysis of bulk powder has already been emphasized. Care must be taken in sampling to ensure the sample also is representative and unbiased with respect to the characteristic being examined.

The number of samples required depends upon the type of material, the purpose of analysis, and the characterization method.

The degree of particle dispersion affects the measured particle size. The measured particle size may be that of individual particles, aggregates, or flocculates. An aggregate is a group of two or more individual particles strongly held together by forces that are stable in normal handling but broken up by shear forces. A flocculate is a group of aggregates or particles in a liquid suspension that break up by normal shaking and stirring but rapidly reform on standing (Herdan et al. 1960e). These are usually characterized by secondary minima interactions according to the Derjaguin-Landau-Verwey-Overbeek (DLVO) theory in aqueous suspensions, and by steric effects in nonaqueous solvents. Particles may be dispersed in a dry or wet state. Shear forces provided by pressure drops across a measurement zone are a common method for dry powder dispersion. The degree of dispersion depends upon the shear forces applied and the interparticulate forces within the aggregates. When particles are dispersed in a liquid medium, the dispersant should not produce dissolution or be chemically reactive with the sample. The addition of surfactants and sonication may be used to aid particle dispersion.

Measurement bias may be a result of incomplete dispersion of aggregates and particle comminution.

Various sizing instruments to determine equivalent spherical diameters are based on different principles. The results obtained by each instrument are rarely in agreement (Herdan et al. 1960e). Particle size may be determined from measurements in one dimension (diameter, width, or length), two dimensions (area), or three dimensions (surface area or volume). All three estimates will be numerically similar for spherical particles. However, the estimates for nonspherical particles depend on particle shape. The size range and degree of size resolution of measurement depends on the measurement principle used. Measurement bias arises if a portion of the measured size distribution falls outside the size range accurately measured by the instrument. Knowledge of the limitations of each instrument and awareness of the systematic error caused by measurement bias enables meaningful interpretation of the observed characteristics of the powder sample (Etzler and Sanderson 1995).

## Microscopy

The sources of error in size measurement by microscopy include sample preparation, dispersion and mounting, optical arrangements, and measurement procedures (Herdan et al. 1960a). The sample must be completely dispersed. Difficulties arise in the size measurement of aggregated or contacting particles due to unclear boundaries of individual particles under these circumstances. Measurement should be carried out on a series of randomly selected fields. All the particles in a complete field should be measured to avoid bias. A smaller number of particles are contained in each field when using microscopes with greater resolution, which makes the sample less representative of the whole. Measurement of an insufficient sample size may be a source of error. Analysis of at least 300 particles is required for representative sampling (Herdan et al. 1960b).

## Sieving

Particle size distribution depends on the sieve loading, duration of sieving, frequency and amplitude of sieving, and random orientation of particles. The passage of a particle through a sieve aperture does not solely depend on its relative size dimensions. Other factors include the size and quantity of surrounding particles, access to the aperture, deviations from the nominal size of the aperture, and the relative orientation compared to the aperture. Comminution of larger particles may allow passage through smaller apertures. Aggregation of small particles may prevent passage through the given aperture. Higher accuracy in sieve analysis is obtained by increasing the sieving duration. Larger loads require longer sieving times for complete separation; however, reducing the load is more effective than increasing the sieving duration. The effect of overloading is more significant for smaller sieve apertures. Repeated sieving of a particular sample with the same equipment and experimental conditions will not give identical results (Herdan et al. 1960b).

## Sedimentation

Particle size in sedimentation is calculated from Stokes' law, which assumes that particles are rigid spheres, the fluid is incompressible, the fluid velocity at the particle surface is zero, and particle motion is constant and not affected by other particles or nearby walls. However, these assumptions are not always met. Corrections are made for nonspherical particles and the effects of the bottom and walls of the container. The motion of neighboring particles may affect the sedimentation rate and calculated particle size (Herdan et al. 1960b). Accurate size determination by sedimentation requires complete dispersion of particles in a suitable liquid medium that is free of air and disturbances. A dispersing agent may be added, and the density and viscosity may be adjusted to suit the particle density and size to ensure that the rate of sedimentation lies within the Stokes region (Herdan et al. 1960c). Accurate temperature control is required to prevent changes in the viscosity of the dispersing media.

## Laser Diffraction

Laser diffraction instruments use either the Fraunhofer or Mie theories to determine particle size from the diffraction pattern of particles. The Fraunhofer theory assumes that the wavelength of light is significantly smaller than the particle diameter, particles of all sizes scatter light with equal efficiency, and the particle is opaque and transmits no light. However, these assumptions do not hold true for many pharmaceutical particles. The Mie theory is more appropriate for particles smaller than 10 μm. Knowledge of the refractive index of the material and dispersing medium is required when using the Mie theory (Etzler and Deanne 1997). Laser diffraction instruments use mathematical algorithms to determine the particle size distribution from the complex diffraction patterns obtained. These algorithms vary between manufacturers and software versions. In addition, they are not disclosed because they are considered proprietary (Etzler and Sanderson 1995).

Caution must be employed when characterizing fine powders with laser diffraction instruments. The particle size distributions obtained from three different laser diffraction instruments employing the Fraunhofer optical model have been shown to differ from each other and from the results obtained from microscopy, TOFABS, and light scattering (Etzler and Sanderson 1995). Inaccuracies were observed in the particle size distributions of small or transparent particles determined by laser diffraction. In addition, small "ghost" particles appeared due to high angle diffraction of the edges of nonspherical particles. Significant differences also have been observed in the particle size distribution obtained from two different laser diffraction instruments employing the Mie optical model (Etzler and Deanne 1997). These differences were attributed to the mathematical algorithms used to analyze the diffraction data. Thus, dissimilar size distributions may be observed even when employing the same measuring principle. When size distributions of identical samples obtained by laser diffraction instruments using the Fraunhofer and Mie optical models were compared, the use of the Mie optical model did not increase the accuracy (Etzler and Deanne 1997).

## Conclusions

Sampling is a selection process used to estimate population characteristics from a small portion of the population. The accuracy of population estimates relies on the sample selected to be representative and nonbiased. The various types of sampling schemes include random, systematic, stratified, and cluster sampling. Various sampling methods and devices for the reduction of bulk powders to laboratory-sized samples to measurement-sized samples are described in this chapter. The use, limitations, and potential sources of bias of these sampling techniques are highlighted. The specific sampling of powder blends, liquid suspensions, and particulate aerosols are described. The causes and mechanisms of segregation are discussed, together with errors involved with sampling, sample preparation, and measurement. Knowledge of the fundamental particle behavior and theory underlying the various sampling methods is important for the understanding of the inherent errors of sampling.

Knowledge of the characteristics and performance of powders is always seen or interpreted through the filter of sampling and statistical probabilities. The pharmaceutical scientist and powder technologist should always review the nature of this filter and remember that a true population parameter exists for which an estimate is all that can be achieved. The goal is to ensure that this estimate is a reliable reflection of the true parameter and is relevant for the desired application.

# 4

# Particle Size Descriptors and Statistics

Particle size analysis has been the subject of a number of excellent texts (e.g., Stockham and Fochtman 1979; Stanley-Wood and Lines 1992; Allen 1990a). Particle size clearly plays a role in formulation development (Piscitelli et al. 1998; Dunbar and Hickey 2000; Vaithiyalingam et al. 2001).

Preparation of particles for use in pharmaceutical products involves a number of different processes. A general scheme may be outlined as shown in Figure 4.1.

The physicochemical properties of the drug or excipient play a role in the adoption of conditions under which each of these processes is conducted (Sacchetti 2000; Tong and Whitesell 1998). The initial solution or suspension characteristics is dictated by the solubility and dissolution of the drug in the selected solvent, which in turn relates to the polymorphic form of the starting material (Wostheinrich and Schmidt 2001; Brittain 1999, 2000). Nucleation is influenced by the presence of impurities and lattice formation. If the

**Figure 4.1** Flow diagram for common processes involved in the preparation of particles suitable for inclusion in pharmaceutical products.

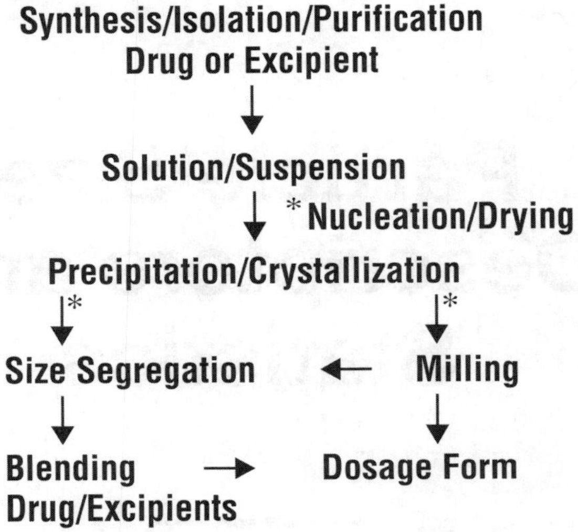


starting particle size is not suitable for preparing a pharmaceutical product, size reduction is required. Milling involves mechanical energy input to reduce the size of particles (Hickey and Ganderton 2001; Mura, Faucci, and Parrini 2001; Suzuki et al. 2001). Fractures in the crystal lattice create new surfaces, and recrystallization phenomena thus may occur during the manufacturing process (Parkkali, Lehto, and Laine 2000; Takeuchi et al. 2000; Darcy and Buckton 1998). Specifications on the particle size required for a product involve a defined size range, which may require active segregation (sedimentation, sieving, electrostatic precipitation) (Allen 1990a). This requires either overcoming the cohesive nature of the particles (Tomas 2001a, 2001b) or the formation of stable aggregates (Pietsch 1991). Once the particles of each of the components (drug and additives) have been produced, they are blended to uniform drug dispersion (Villalobos-Hernandez and Villafuerte-Robles 2001; Venables and Wells 2001; Michoel, Rombaut, and Verhoye 2002; Hwang,

Gemoules, and Ramlose 1998). Segregation and blending methods may result in some attrition of particles, and changes in particle size may occur.

Particle size plays an important role in each of these processes. In addition, stable aggregate size may play a role in the performance of the product. In this case, the forces of interaction between particles must be considered (Rietma 1991a). These include ionic, covalent, charge / complexation, van der Waals, capillary, and mechanical interlocking forces (Israelachvili 1991c). Particle interaction influences aggregation (Pietsch 1991), flow, dispersion (Tomas 2001a, 2001b; Yi et al. 2001; Kano, Shimosaka, and Hidaka 1997), compression (Hwang, Gemoules, and Ramlose 1998; Garekani et al. 2001b), moisture content, sampling (Thompson 1992b), and statistical descriptors of powders.

Ultimately, distribution of particles in a blend impacts upon the quality criteria for the unit dose (Orr 1981). The amount of drug a patient receives in the dosage form (e.g., tablets, capsules), should reflect the label claim for that product. Official specifications cannot simply be the subject of regulations independent of science and technology. Quality control can be maintained only with a basic understanding of the pharmaceutical technology involved in manufacturing.

Variations in the amount of drug that a patient receives from a single dose depend on two factors, actual volume or weight of therapeutic agent and degree of homogeneity obtained as a result of the mixing processes. Homogeneity in this context is something of a misnomer. The homogeneity of a blend or suspension, unlike the molecular dispersion of a solution, is limited by the degree of subdivision of the drug particles. The distribution of drug particles within the diluent dictates uniformity. Particle behavior, also a function of particle size, may be a factor in uniformity. A suspension, for example, may be more or less stable based on particle size and density with respect to the dispersion medium. The ability to redisperse the drug particles following storage may also be influenced by their characteristics.

The influence of particle size and distribution on pharmaceutical processes and ultimately on the efficacy of drugs delivered from solid dosage forms cannot be discussed without considering the methods for describing these parameters. Particles of any substance exhibit unique morphologies. These morphologies are usually related to the chemical composition of the particles and the method of manufacture and processing (Hickey and Ganderton 2001). This chapter will not consider the ways in which different particles are obtained, which was addressed in earlier chapters; rather, it will focus on the way the particle morphology contributes to estimates of particle size.

Real particle populations consist of heterogeneous mixtures. Individually classifying particles by size, from smallest to largest, may result in a shift in shape, surface rugosity, and energetics. Since these characteristics may play important roles in particle performance, we may infer that a good understanding of particle size and distribution is important to process development.

Some methods of manufacture, such as spray drying, result in spherical particles (Sacchetti and van Oort 1996; Palmieri et al. 2001; Kumar, Kang, and Yang 2001). Particles with spherical morphology and smooth, nonporous surfaces are easily characterized because their diameters unambiguously describe their sizes (Stockham and Fochtman 1979). This ease of characterization derives from the infinite rotational symmetry associated with these particles. Simply stated, it does not matter how you measure the size of these particles or who conducts the study. Within the limits of error of the scale of measurement, the same answer will be obtained under all circumstances. For the moment, this ignores the issue of sampling (Thompson 1992b). As noted in earlier sections, however, most pharmaceutically relevant solids are crystalline in nature and exhibit a range of crystal systems and habits, resulting in a variety of shapes and surface irregularities (Carstensen 1993).

## Particle Size Descriptors

In order to communicate clearly the particle size and distribution of irregularly shaped particles, a "language" of descriptors must be agreed upon by those working in the field. This raises the question, "How many different ways can we consider the dimensions of a particle?" The best way to address this question is to take into account the intended purpose of the particles.

## Particle Shape

Irregular particles are difficult to reproduce accurately in terms of their surface invaginations. Sophisticated attempts to describe particle peripheries have been attempted by using polar coordinates from which Fourier coefficients (Luerkens 1991) may be derived or by using a fractal approach (Kaye 1989). However, these methods are tedious and lack broad application. Other attempts at irregular particle shape description involve Heywood and Waddell shape factors, which are described in Chapter 5 (Fayed and Otten 1984).

## Particle Size

The convention for describing particle size has been to consider irregular particles in terms of spherically equivalent particles with the same surface area, volume, and projected area, which are static measures, or with the same sedimentation rate and aerodynamic behavior, which are dynamic measures. The reason for considering each of these equivalent parameters relates to the application of the data. For example, surface area may be important for catalysis, whereas sedimentation may be important for the stability of a suspension formulation.

The particle size descriptor for individual particle sizes can be traced back to the method by which it was determined. The following brief review of some key particle size descriptors is discussed in more detail in Chapters 5, 6, and 8.

Microscopic analysis of particles has resulted in the determination of a variety of particle size dimensions, including Feret's diameter, Martin's statistical diameter, and Heywood's projected area diameter (Reimschuessel, Macur, and Marti 1988; Amidon 1992). Feret's diameter may be defined as the distance between two tangents drawn at the periphery of the particle perpendicular to the direction of scan. Invoking the direction of scan was anticipated to allow accurate mean data to be derived because variations in shape of particles would be averaged by their random orientation. However, this dimension frequently overestimates the population mean size. Martin's statistical diameter involves bisecting a particle by area in the plane of scan and then measuring the length of the line as it crosses the particle periphery. Ultimately, the measure that found most favor for measuring the geometric size of particles from visual observation was Heywood's projected area diameter, or simply the projected area diameter. This measure involves equating the area of a circular disc to the projected area of the particle and using the diameter of the disc as the characteristic dimension of the particle. Figure 4.2 illustrates the diameters for comparison.

To simplify the estimation of projected area diameter, graticules have been developed for light microscopy, which show discs in $\sqrt{2}$ scales from which sizes can be estimated directly. The most commonly used graticules are the Patterson Globe and Circle or Porton graticules (Green and Lane 1964). An important factor in this estimation is that particles being examined microscopically are generally in the plane of greatest stability, which means the third dimension in space, depth, is not being assessed. For plates and fibers, this may lead to misleading estimates of particle size.

Sedimentation of particles is known to be a function of Stokes' law in that particles are subjected to gravitational acceleration (g) and resisted by the viscosity (η) of the medium through which they

**Figure 4.2** Characteristic particle diameters: (A) projected area, (B) Martin's statistical, (C) Feret's.

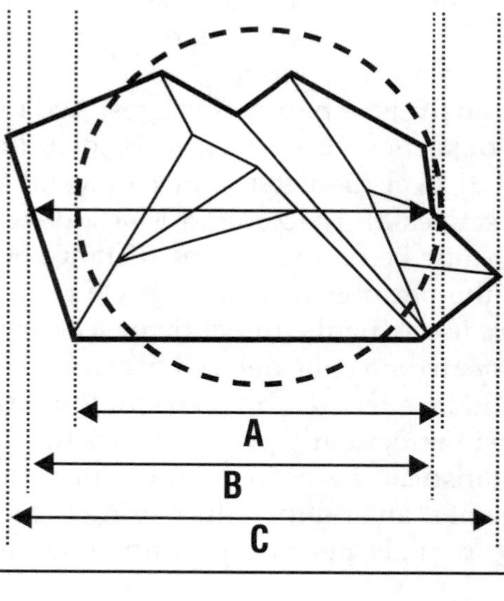

are falling. The general expression of Stokes' law governing this phenomenon is as follows:

$$\frac{h}{t} = \frac{\rho d_{st}^2 g}{18\eta} \tag{4.1}$$

where t is the sedimentation time, h is the height through which the particle falls, $d_{st}$ is the Stokes diameter of the particle, and ρ is the difference in density between the particle and the medium.

Consequently, the Stokes diameter can be derived from the height through which a particle will fall as a function of its size and time of sampling. This is a particularly important parameter for suspension stability and may be used to select bulking agents and to design processes that require suspended particles.

A variety of individual particle sizes are based on other geometric properties of particles, the most notable among these being surface area and volume. The surface area of a sphere is related to the diameter by the term $\pi d^2$, and the volume is related by $(\pi d^3)/6$. Methods such as gas adsorption surface area measurement and electrical resistance may be employed to estimate the equivalent surface area and volume diameters, respectively, of particles being studied (Grant 1993).

Caution should be taken not to convert particle size data based on one set of particle properties to another, which can lead to significant errors. Wherever possible, the particle sizing technique should duplicate the process under evaluation.

## Powder Mixing

In the foregoing section, particle size was viewed in terms of evaluating the individual particle size from which a population of particle sizes can ultimately be constructed. The degree of homogeneity of a powder blend is a key area in which particle size and distribution play a role (Venables and Wells 2001). Two major processes for powder mixing—dry mixing and solvent—are shown in Figure 4.3.

Dry mixing requires control of particle size and distribution, enabling prediction of content uniformity and dissolution rate. Flammable solvents are not required in this method, so the safety hazard

**Figure 4.3** Principal methods for formulating tablets containing small amounts (<5%) of drug.

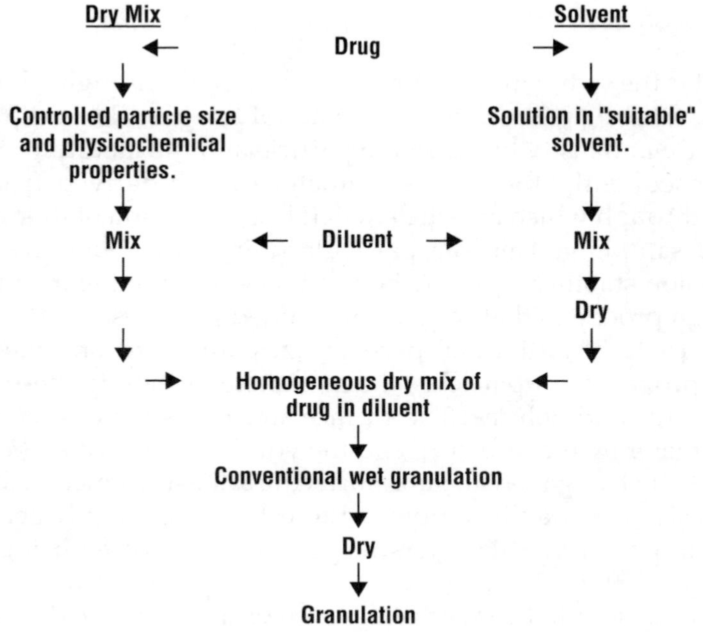


is reduced. Homogeneity may be difficult to achieve if fine particles are employed, however, because they may aggregate, causing poor dispersion of the drug. The solvent method theoretically results in a more ordered mixing and dissolution rate. However, the final particle size and physicochemical properties of the product are difficult to predict. Among the possible undesirable outcomes from the application of the solvent method are solvate formation, residual solvent, degradation during processing, and variation in crystal conditions. Solute migration may give rise to drug-rich fractions. In addition, dry mixing is essential subsequent to the solvent method, and flammability may be a hazard during processing. In either case, and for subsequent granulation, a knowledge of particle size and distribution of all the components is important in establishing processing parameters (e.g., mixing; flow and dispersion; the pharmaceutical characteristics, particularly content uniformity and dissolution rate; and specifications on the quality of the final product).

## Particle Size Statistics

"Real" or "true" distributions of particles cannot be known due to shortfalls in our experimental approach as well as an inability to capture the entire nature of a particle population. The two general approaches to evaluating particle size are (1) measuring derived properties of individual particles and summing them for a distribution; and (2) observing a physical phenomenon that occurs for a group of particles (Goldman and Lewis 1984; Reist 1993b). Even when used correctly, these techniques are subject to error. In extreme cases, abuses of the methods can lead to misleading information.

### Principles of Statistical Inference

The sequence of events in describing particle size distributions begins with the selection of a particle measurement technique. Once this technique is selected and applied, the sequence of events in evaluation of the data is as shown in Figure 4.4.

Data collected in an appropriate manner form the basis for statistical evaluation; therefore, the production of data must be planned carefully. A poorly planned experiment or measurement does not lend itself to statistical analysis. In defining a problem, a number of issues may be considered (Taylor 1990). Figure 4.5 outlines these

**Figure 4.4** Sequence of steps involved in evaluating particle size distributions [modified from Taylor (1990)].

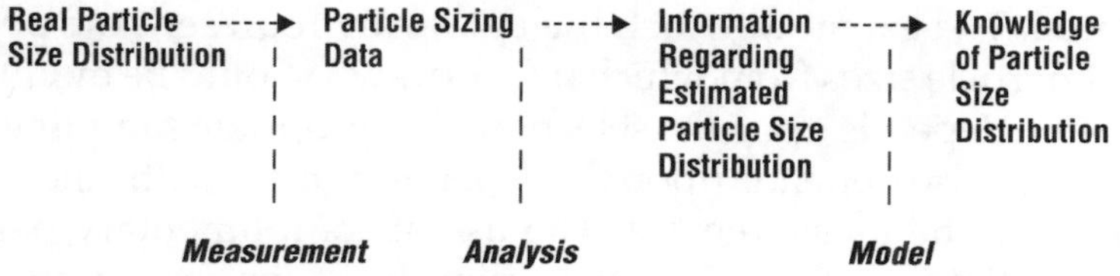

**Figure 4.5** Defining the purpose of the particle size data to be collected.

**Desired outcome of measurement (application)**
**Particle population**
**Paramenter to be measured (relevance to application)**
**Knowledge of the problem (application)/ particle population (facts)**
**Assumptions made to initiate measurement (about particles or limitations of measurement technique)**
- **Basic nature of problem (application)**
  - **Research and development**
  - **Quality control**
  - **Regulatory compliance**
- **Temporal nature of problem**
  - **Long-term**
  - **Short-term**
  - **Single study**
- **Spatial nature of problem**
  - **Global**
  - **Local**
- **Prior art**
- **Other factors**

considerations. It is clear from this list that measurements are not conducted in an information void. Everything that may be known about the system under investigation must be employed to devise, or select, an appropriate measurement technique and procedures for the evaluation of particle size and distribution.

The collection of certain indicators of the quality of data is important. First, the quantitative value of the data must be evaluated. This requires validated methods that have known limits of detection and sensitivity under defined experimental conditions. Quantitative limits for the accuracy and precision of a technique should be stated. The extent to which the data truly represent the population of particle sizes from which it was drawn should be monitored. This often depends upon the selection of appropriate sampling techniques, but also depends upon the expert judgment of the individual conducting the measurement. The use of complimentary particle-sizing methods will also impact upon decisions regarding representative sampling.

It is important that the data collected are complete. Certain particle sizing techniques may not measure the entire distribution, which becomes an important factor in the interpretation of the data. A complete analysis of the data will include acknowledgment of the limitations to the methods involved and their impact on the data interpretation.

Further, it is important that the data collected facilitate comparison. This comparison can be made only within a particular method of particle size analysis. Comparisons between differing methods are usually complimentary but are not expected to give numerically similar results. All of the issues discussed in the foregoing section become relevant in the use of particle sizing methods as quality control tools (Shewhart 1986; Deming 1964).

Errors and uncertainty occur at each step described in Figure 4.5. The data collected should be factual— that is, truly representative of a physical reality— or statistical analysis will be meaningless. The errors that can occur in particle size measurement are similar to those that would be expected in any other experimental situation. These errors may be systematic, random, or due to mistakes.

A systematic error is always of the same magnitude and sign and is usually viewed as a bias. Wherever possible, the study should be designed to avoid bias. However, circumstances may occur in which the bias is inherent in the method; such bias should then be acknowledged and alternative methods considered. Indeed, this was the case with early methods of estimating particle size by using microscopy in which Feret's diameter, for example, was known to overestimate the size of particles. Ultimately, the projected area diameter became the more frequently used descriptor of particle sizes obtained by microscopy. Systematic errors may also occur during data analysis and should be given equal consideration when manipulating data for interpretation.

Random errors occur in any experimental work and are the basis for setting standards on the limits of accuracy and precision. These are handled by the variance considered in statistical analyses.

Mistakes, which may occur in the collection of data or in its analysis and interpretation, do not lend themselves to statistical analysis. A measurement system that is unstable and not in statistical control cannot produce meaningful data (Taylor 1990).

## Mathematical Distributions

The description of particle populations requires adoption of mathematical descriptors of the entire distribution of sizes. Statistical or other mathematical principles have been employed to fit particle size data in an attempt to describe them succinctly.

It is enormously tempting to fit particle size data to a normal or Gaussian distribution of the form:

$$y = \frac{1}{\sigma\sqrt{2\pi}} e^{-(x-\mu)^2/2\sigma^2} \tag{4.2}$$

where $\mu$ is the arithmetic mean and $\sigma$ is the root mean square (RMS) deviation of individual sizes from the average. However, experimental data rarely conform to a normal distribution. Considering the method by which most particles are formed, a normal distribution of particle sizes would not be anticipated. Milling and spraying procedures used to prepare particles are destructive methods that result in the production of small particles from large ones. This is a phenomenon dominated by the volume of particles involved. Since this is a volume function, in broad terms there is a cubic relationship between the large numbers of small particles that can be obtained from small numbers of large particles. However, any method of particle production results in a range of sizes that will not conform strictly to a cubic relationship. Consequently, a number of statistical and mathematical functions have been used to fit particle size distribution data. The range of distributions that may be considered include normal, lognormal, power functions, Rosin Rammler, modified Rosin Rammler, and Nukiyama-Tanasawa (Dunbar and Hickey 2000).

Lognormal distributions are derived from the normal distribution of the logarithms of the particle sizes. Early evaluations of particle size statistics concluded that many populations of particle sizes could be approximated by lognormal mathematical functions. Indeed, the collapse of lognormal distributions into normal distributions was used as a measure of the degree of dispersity of a population of particles. The point of collapse is used to define the practical limit to monodispersity (equivalent to a geometric standard deviation of 1.22, Fuchs and Sutugin 1966).

The advantage of normal and lognormal distributions is that they may be used to extrapolate between different measures of distribution, for example, mass or number. Hatch-Choate equations may be used for this purpose in the case of lognormal distributions. The frequency term for the distribution should be distinguished from individual size parameters. The entire population of particles can be divided into sizes according to specific features such as volume, surface, number, and so forth.

## Multimodal Distributions

Complex phenomena may occur based on additive particle size distributions in which multimodal distributions occur. Multimodal distributions require deconvolution if the contributing distributions are to be characterized (Esmen and Johnson 2001). If broad, additive distribution occurs, then the overall distribution would appear to fit a normal function according to the central limit theorem (CLT).

Assuming the data have been collected by using a validated method, they may still be easily misinterpreted. Data interpretation depends on the researcher's intent and understanding of the method employed. The first area in which misinterpretation can occur relates to the limitations of the technique used for determining the particle size. Assumptions are often made to conduct an analysis that must be known to allow interpretation of the data. The following list itemizes key examples:

- The ability to examine the entire distribution of particles is a significant requirement. If the distribution extends beyond the working range of the instrument, this fact should be stated as part of the data interpretation. Ideally, the method will be capable of evaluating the entire distribution.

- The appropriate use of a method is a requirement. If equivalent volume diameters are obtained by laser diffraction, which is an optical method, but the application is to suspensions of particles in a liquid dispersion medium, the data may be of limited relevance to the application. In this example the Stokes diameter would be most relevant to the application.

- A method that meets the first two criteria may be selected, but sample size may not lend itself to statistical analysis. This is a frequently debated topic with regard to particle sizes derived from microscopic examination, in which the prospect of measuring thousands of particles can be intimidating.
- Finally, the application of a mathematical distribution to fit the practical distribution must be subjected to great scrutiny. The closeness of the fit and, if possible, a physical model to explain the rationale for the fit are desirable. In the early part of the twentieth century, a major reason for searching for mathematical fits to the data was to allow quantitatively meaningful communication among researchers that would allow studies to be reproduced scientifically or would allow specifications on a product to be set. This should be considered in the historical perspective, in which raw data presented even graphically could be cumbersome. In the early part of the twenty-first century, we are truly in the information revolution in which large quantities of data can be routinely handled with the computer capacity that most people have on their office desks. It is therefore no longer as important to fit mathematical distributions to practical data to convey this information easily. Indeed, it is arguably more desirable to give the raw data, without interpretation, to those who wish to reproduce findings or set standards. The importance of mathematical analysis is perhaps most relevant to understanding the principles of particle formation and process reproducibility, in which case the best fit to the data is more important than a single model that fits all particle size distributions. Certainly, the application of the lognormal distribution has been stretched beyond its ability to accurately describe many particle populations. However, presenting raw data is not necessarily as simple a task as it first appears. Some instruments employ algorithms that may not allow optimal visualization of data (Lawless 2001). Indeed, the method that the individual scientist may employ to graphically represent data may be misleading (Menon and Nerella 2001.) Therefore, caution should be exercised in mathematical, statistical, or graphical representation of the data.

The way forward in presenting data for pharmaceutical purposes lies in the use of complimentary techniques. Direct microscopic images of particles help in the interpretation of particle sizing data obtained by other methods. The shape of the particle size distribution and an effective median diameter defined according to the application of the data should be obtained. Where the data application predominantly relates to weight, volume, number, or surface area and is unique for a pharmaceutical process or parameter, the equivalent median diameter (e.g., equivalent weight median diameter for content uniformity, equivalent weighted surface area diameter for dissolution, equivalent aerodynamic diameter for respiratory deposition) may be weighted in two ways—for the process and for efficiency. These diameters could be suitable for defining specifications and for monitoring the effect of particle size on quality parameters of medicines in development, pharmaceutics, efficacy, or clinical investigations.

It is absolutely essential that particle size data be presented uniquely for each pharmaceutical parameter it is controlling. An inhalation product, for example, has at least three separate reasons for controlling particle size: lung deposition, for dissolution upon deposition, and content uniformity and mixing aspects in the initial formulation. In this situation, the overriding factor is lung deposition, but separate calculations need to be performed for the other parameters to understand the effect of working at the bottom and top end of the aerodynamic particle size specification on the content uniformity and dissolution. Even given this analysis, the interpretation does not allow for processing aspects related to agglomeration of fine particles.

## Applications

The assumption of monosized spherical drug and excipient particles in a free-flowing powder for tableting is one of the simplest examples of the application of particle size analysis to the preparation of a dosage form. A random mix can be achieved when the drug and excipient have reached a stable maximum distribution. This is a mathematically random mix because probability is the only factor involved in dispersion. For a completely randomized mix of spherical particles, for visualization purposes differing only in color, the standard deviation, $\sigma_R$, has been given by

$$\sigma_R = \sqrt{\frac{XY}{n}} \tag{4.3}$$

where X and Y are the proportions of drug and excipient particles, respectively, and n is the total number of particles in the tablet. For a completely unmixed system, the standard deviation, $\sigma_0$, is given by

$$\sigma_0 = \sqrt{\frac{X}{Y}} \tag{4.4}$$

A more complex model is required, however, because this is a highly idealized view. Equation (4.5) uses a binomial distribution of particle sizes:

$$\sigma_R = \frac{XY/M}{Y_w(1+s^2/w^2)_x + X_w(1+s^2/w^2)_y} \tag{4.5}$$

where X is the fraction by weight of the minor component in the mixture (i.e., the fraction of the drug in a tablet); Y is the fraction by weight of the major component in the mixture (i.e., the fraction of the excipient in a tablet); M is the mass of the sample taken from the mixture (i.e., the weight of a tablet); $w(w_x, w_y)$ is the mean particle weight of the mix components; and $s(s_x, s_y)$ is the standard deviation of the particle weight of the mix components.

Equation (4.3) can be modified, and ultimately equation (4.5) can be rewritten as follows:

$$\sigma_R^2 = \frac{XY/M}{Y\Sigma(fm)_x + X\Sigma(fm)_y} \tag{4.6}$$

where $\Sigma(fm)$ is the mean effective particle weight of the component, which is related to $d_{vv}$, the volume-weighted, volume mean diameter, as follows:

$$\Sigma(fm) = \frac{1}{6}\pi d_{vv}^3 Q \tag{4.7}$$

The coefficient of variation, $C_R$, for a random mix is given by

$$C_R = \frac{\sigma_R}{X} \tag{4.8}$$

Expression (4.9) is developed by using the Poisson distribution and regarding the variance due to the particle size distribution as a linear combination of several independent randomly distributed variables:

$$C_R = \frac{100\sigma}{X} = 100\sqrt{\frac{\Sigma(fm)_x}{M_x}} \tag{4.9}$$

Use of the Poisson distribution assumes that the random selection of a drug particle as opposed to an excipient particle is a rare event, and is satisfactory only when the concentration of the drug is low (≤1 percent by weight) and the size of the excipient particles is equivalent to or larger than the size of the drug particles. Substituting the volume-weighted, volume mean diameter, $d_{vv}$, into the expression for the coefficient of variation (4.9) yields

$$C_R = 100\sqrt{\frac{\frac{1}{6}\pi d_{vv}^3}{M_x}} \tag{4.10}$$

Taking the logarithm of each side of equation (4.10) shows that for a tablet formulation containing a low concentration of drug, there will be a linear relationship between log $C_R$ and log $d_{vv}$, as shown in Figure 4.6.

This relationship is useful in determining the maximum particle size of a drug during formulation design. However, the maximum particle size of a drug can be specified only if a decision is made as to the limiting value of $C_R$, the coefficient of variation for a random mix. The decision is to some extent arbitrary, and Johnson (1972) recommends a value of $C_R$ of 1 percent or less.

## Conclusion

The degree of subdivision of a powder impacts on dispersion and sedimentation of suspensions, flow, mixing, and dispersion and dissolution of solid particles. In the case of aerosol products, particle size and distribution also play roles in dispersion, aerodynamic behavior, and deposition.

**Figure 4.6** Theoretical relation between particle size of drug (volume weighted, volume mean diameter, $d_{vv}$) and content uniformity of tablets (coefficient of variation for a random mix, $C_R$).

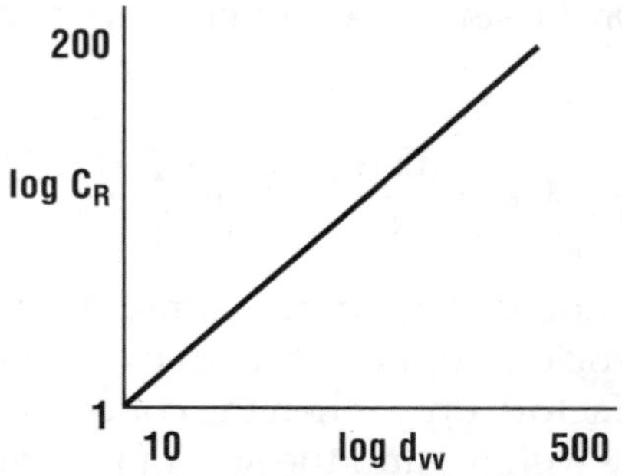


The characteristic dimension of a particle requires definitions of terms. Since most particles are irregularly shaped, the common practice is to consider equivalent spherical dimensions based on a particular feature such as volume, surface area, or projected area diameters. Populations of particles require numerical data analyses. These data may be represented graphically to convey the particle size distribution. Furthermore, the distribution may be interpreted by statistical or mathematical functions to fit the data. The most frequently employed fit to the particle size distribution is the lognormal function. The parameters derived from this fit are the median diameter and the geometric standard deviation. As a cautionary note, mathematical and statistical functions in their simplest forms may not accurately approximate the data. Indeed, sophisticated approaches are required for multimodal data analysis.

Wherever possible, particle sizing data should be matched to its application. For example, the Stokes diameter, based on sedimentation, is of most relevance to suspensions, and the aerodynamic diameter, based on airborne behavior, is of importance for the deposition of aerosol particles.

The remaining chapters discuss the techniques for particle size measurement and the rationale for their selection.

# 5

# Behavior of Particles

Particle behavior is directly related to particle properties, including size, distribution, density, shape, and hygroscopicity. Particulate systems exist as bulk powders or particle dispersions in liquid or gaseous systems. Particle behavior within these systems is greatly dependent on the forces between particles and the effect of external forces on particles. Fundamental particle properties, such as particle size, shape, surface roughness, surface area, and density affect particle interactions; and have a significant effect on particle behavior. Particle interactions involve electrostatic forces, intermolecular forces, capillary forces, and solid bridging forces. Methods of determining adhesion force include single and multiple particle detachment. This chapter describes the nature of particle motion within bulk powders and particle dispersions in gaseous and liquid media, including the flow, fluidization, and mixing of powders, as well as diffusion and gravitational sedimentation of dispersed systems.

## Physical Properties

Powder properties may be classified as fundamental or derived. Fundamental properties are inherent attributes relating to individual particles or particle populations, such as size and size distribution, shape, porosity, true density, and surface area. Derived properties depend upon the fundamental properties and may be affected by environmental conditions. Examples of derived properties include adhesion and flow.

## Size

The size of a sphere is uniquely defined by its diameter. However, real particles are rarely homogeneously spherical in nature. The equivalent spherical diameter relates the size of an irregularly shaped particle to the diameter of a sphere having the same size-dependent physical property (Allen 1975; Martin 1993e). Typical physical properties measured include volume, surface area, projected area, and settling velocity.

Particle sizing techniques may be classified as direct imaging, indirect imaging, physical separation, and charge analysis methods. These methods and instrumentation are described in detail in Chapter 6. The equivalent diameter measured by using one method is not directly comparable to sizes measured by using other methods due to the fundamental principles involved in each measurement technique. Thus, the sizing method chosen should match the particle behavior required.

A population of pharmaceutical particles rarely contains uniformly sized (monodisperse) particles. Polydisperse particle populations generally fit a lognormal size distribution and are characterized by the geometric mean size and geometric standard deviation. Statistical analysis of particle populations are discussed further in Chapter 4.

Particle size is a very important parameter for predicting the behavior of powder systems. Particle size greatly affects many derived particle properties, such as particle adhesion, flow, and fluidization. The fundamental laws of particle motion are strongly governed by particle size.

## Shape

Pharmaceutical particles are rarely spherical in nature. The description of particle shape requires the use of dimensionless numbers that reflect the magnitude of particle irregularity. Shape factors are used to describe particle shape as a deviation from regular geometric shapes, such as spheres and cubes. Other approaches are based on mathematical methods, such as Fourier and fractal analyses.

### Shape Factors

Many different types of shape factors exist. The Waddell sphericity factor ($\psi$) is the ratio of the surface area of a sphere with the same volume as the particle to the surface area of the particle (Dallavalle 1948g). For a sphere, then, $\psi = 1$, and as particles deviate from sphericity, $\psi < 1$. Similarly, the Waddell circularity factor (¢) is the ratio of the circumference of a circle with the same cross-sectional area as the particle to the perimeter of the particle (Dallavalle 1948g). The volume and surface shape factors relate the equivalent diameter to its particle characteristic (Dallavalle 1948g):

$$V = \alpha_v d_v^3 \tag{5.1}$$

and

$$S = \alpha_s d_s^2 \tag{5.2}$$

where V is the volume of the particle; $d_v$ is the equivalent volume diameter; $\alpha_v$ is the volume shape factor; S is the surface area of the particle; $d_s$ is the equivalent surface area diameter; and $\alpha_s$ is the surface shape factor. The surface-volume shape coefficient ($\alpha_s/\alpha_v$) is the ratio of the surface to volume shape factors (Allen 1975). The value of $\alpha_s/\alpha_v$ equals 6 for spherical particles and exceeds 6 with more asymmetrical particles.

The specific shape factor, ($\alpha_{sv}$, relates particle shape to surface area (Dallavalle 1948g):

$$S = \alpha_{sv}\left(\frac{\omega}{N\rho}\right)^{\frac{2}{3}} \tag{5.3}$$

where S is the surface area; $\omega$ is the weight of particles; N is the number of particles; and $\rho$ is the particle density. For spheres, $\alpha_{sv} = 4.85$, and for cubes, $\alpha_{sv} = 6$. For irregular particles, $\alpha_{sv}$ is greater than 6.

The dynamic shape factor, $\chi$, relates the equivalent volume diameter to the Stokes diameter (Hesketh 1986a):

$$\chi = \left(\frac{d_v}{d_s}\right)^2 \tag{5.4}$$

The value of $\chi$ is 1 for spherical particles and greater than 1 for nonspherical particles.

Other shape factors include the following: Heywood's ratio described particle shape by using the three mutually perpendicular dimensions of length (L), breadth (B), and thickness (T) (Allen 1975). The elongation ratio is the ratio between the length and breadth. The flakiness or flatness ratio is the ratio between the breadth and thickness. The area ratio ($\alpha_a$) and prismoidal ratio ($p_r$) are related to the projected area and volume of the particle, respectively. The values of $\alpha_a$ and $p_r$ enable classification into tetrahedral, prismoidal, subangular, and rounded.

Another shape factor, NS, is a composite shape factor for two-dimensional particle outlines that considers the deviation from circular, square, and triangular geometric shapes, particle elongation, and the number of corners (Podczeck 1997). The NS shape enables classification for a range of different shapes.

### *Fractal Analysis*

Fractal analysis is a mathematical tool used to describe the surface morphology and the degree of surface irregularity of a particle as a single parameter, the fractal dimension. The fractal dimension may be determined from either the perimeter or the area of a two-dimensional profile. The concept of fractal geometries is based upon self-similarity at any dimension and relies on repetitive patterns at any scale of scrutiny.

For fractal analysis based on particle perimeter, the length of the perimeter depends on the stride length used to measure it. The smaller the stride length, the larger the perimeter measured. A linear plot (known as a Richardson plot) is obtained by graphing the logarithm of the length of the perimeter against the logarithm of the stride length used for the measurements. The fractal dimension ($\delta$) is determined by using the relation

$$\delta = 1 + |m| \tag{5.5}$$

where m is the slope of the curve of the Richardson plot. The fractal dimension is a description of the space-filling capacity of the profile boundary. Values of $\delta$ generally lie between 1 and 2. The higher the value of $\delta$, the more rugged the boundary.

Fractal analysis of the profile of fine particles may produce two linear regions in the Richardson plot. This is due to the different scales of scrutiny to examine the particle. The fractal dimension obtained by using large stride lengths (low resolution) represents the global spatial structure of the particle, whereas the fractal dimension obtained by using small stride lengths (high resolution) represents the detailed textural structure (Kaye 1984).

The use of fractal analysis for characterization of shape in pharmaceutical particles requires the capture of the silhouette boundary profile of the solid particles, generally by SEM or image analysis. The image needs to be digitized, and touching particles must be separated with the aid of algorithms before the fractal dimension can be determined. Thus, fractal analysis of individual particles can be tedious and time consuming. Nevertheless, fractal analysis has been used to study the surface geometry of pharmaceutical particles, including excipients (Thibert, Akbarieh, and Tawashi 1988), diclofenac-hydroxyethylpirrolidine (Fernandez-Hervas et al. 1994) and $\beta$-cyclodextrin-indomethacin (Fini et al. 1997).

### Fourier Analysis

Fourier analysis provides a measure of particle shape by using a two-dimensional profile. Particle shape is defined here as the pattern of all the points on the boundary or surface of the particle profile (Luerkens, Beddow, and Vetter 1984). The profile of the particle may be defined as a unique set of x,y coordinates that may be used to regenerate the original two-dimensional shape. Fourier analysis is a mathematical approach used to convert the particle profile into a set of mathematical descriptors. The periodic function for all x values that fits the Fourier equation is the following (Hundal et al. 1997):

$$f(x) = a_0 + \sum_{n=1}^{\infty}\left[a_n \cos(nx) + b_n \sin(nx)\right] = a_0 + \sum_{n=1}^{\infty} A_n \cos(nx - \alpha_n) \qquad (5.6)$$

where
$$A_n = \left(a_n^2 + b_n^2\right)^{\frac{1}{2}} \tag{5.7}$$

and
$$\alpha_n = \tan^{-1}\left(\frac{b_n}{a_n}\right) \tag{5.8}$$

The variable n is the harmonic order, and the Fourier coefficients $A_n$ and $\alpha_n$ are the harmonic amplitude and harmonic phase angle, respectively.

Three different methods of Fourier analysis are used to characterize particle profiles (Luerkens, Beddow, and Vetter 1994). The radial method converts the particle profile to polar coordinates of the radial length from the center to a boundary as a function of angle (R,θ). However, this method is not applicable for concave-shaped particles (Hundal et al. 1997). The Granlund descriptors are obtained from the boundary coordinates in the complex plane. This method is widely used to describe particle shapes. The increased elongation and squareness of disodium cromoglycate particles following the treatment with fatty acids has been quantified with numerical values by using this method (Hickey and Concessio 1997). The Zahn-Roskies (ZR) descriptors are obtained from the angular bend as a function of arc length. The ZR method classifies particles into five shape categories (perfect square, square-shaped, elongated, rounded, and very irregular) (Hundal et al. 1997).

## Surface Roughness

Surface roughness of particles may be characterized by using various types of profilometry, including stylus tip profilometry, laser profilometry, and atomic force microscopy (AFM) (Podczeck 1998).

In stylus tip profilometry, a fine, sharply pointed stylus tip, diamond or sapphire, is drawn across the sample surface. The vertical movement of the tip as it traces the surface irregularities provides the height of the surface. The resolution of measurement is dictated by the size of the tip and the aperture angle. The standard tip size is around 5 μm, which enables measurement of similar-sized valleys. Surface roughness may be examined as contour lines across the sample, or as three-dimensional areas consisting of several lines.

However, during the tip motion, soft surfaces may be damaged and larger asperities may be removed. AFM provides higher-resolution three-dimensional images, in the nanometer range, by using smaller tip sizes. Using AFM, the surface topography is scanned with the sharp tip, similar to stylus tip profilometry. No surface treatment or coating is required, and samples may be imaged in ambient air or within a liquid environment (Shakesheff et al. 1996). The use of a "tapping mode," in which the tip is moved up and down between individual sites, enables the examination of soft samples. Due to vertical range limitations of AFM, difficulty arises in measurement of height variations greater than 10 μm. Laser profilometry provides a noncontact method by which the sample surface is tracked by using a laser beam with a spot size of around 0.6–1 μm. The relative depth of the beam spot at the sample surface is determined by using two-beam interferometry.

Surface roughness parameters include rugosity, root mean square deviation, and maximum peak to valley height. The rugosity ($R_a$), also known as the center line average (CLA) height, is the arithmetic mean average distance of all points of the profile from the center line, given by (Podczeck 1998):

$$R_a = \frac{1}{n}\sum_{i=1}^{n}|y_i| \tag{5.9}$$

where n is the number of points and $y_i$ is the distance of the point i from the center line.

The root mean square (RMS or $R_q$) deviation describes the variability of the profile from the center line (Podczeck 1998):

$$R_q = \sqrt{\frac{1}{n}\sum_{i=1}^{n}y_i^2} \tag{5.10}$$

The maximum peak-to-valley height is the difference between the maximum and minimum points of the profile.

Fractal analysis may be used to examine the particle profile determined by profilometry, enabling the surface irregularity to be characterized by using the fractal dimension.

## Surface Area

The specific surface area is the surface area per gram of the sample. The surface area of a powder sample may be determined by either the adsorption method or the permeation method. The adsorption method determines surface area by the amount of gas or liquid that forms a monolayer on the sample surface. The permeation method determines surface area by the rate at which a gas or liquid permeates a powder bed.

The degree of gas adsorption by a solid depends upon the chemical nature of the solid and gas molecules, the temperature, and the partial pressure of the gas. Adsorption isotherms plot the amount of gas adsorbed on a surface as a function of gas pressure. Various types of adsorption isotherms exist (Martin 1993e). Type I isotherms, described by the Langmuir equation, are linear when the logarithm of the amount of gas adsorbed is plotted with the logarithm of gas pressure. Type II isotherms are sigmoidal-shaped isotherms, in which a monolayer is formed by the first inflection, and multilayer formation occurs with continued adsorption. Type IV isotherms are produced by gas adsorption onto porous solids. Isotherms of Types III and V are not common and are produced when the heat of adsorption in the monolayer is less than the heat of condensation on subsequent multilayers. The size of surface adsorption sites can be probed by different sized gas molecules and described in terms of a fractal dimension (Avnir and Jaroniec 1989).

The most common adsorbing gas for surface area determination is nitrogen, and the adsorption isotherm is generally Type II, as described by the Brunauer, Emmett, and Teller (BET) equation (Dallavalle 1984c):

$$\frac{P}{Q(P_0 - P)} = \frac{1}{Q_m C} + \frac{(C-1)}{Q_m C} \cdot \frac{P}{P_0} \tag{5.11}$$

where Q is the volume of nitrogen adsorbed, P is the pressure, $P_0$ is the saturation vapor pressure of liquid nitrogen, $Q_m$ is the volume of gas required to form a monolayer at normal temperature and pressure, and C is a constant related to the difference between the heat of vaporization of a monolayer and the heat of vaporization of liquid nitrogen.

A linear plot is obtained by plotting $P/[Q(P_0 - P)]$ against $P/P_0$, which allows $Q_m$ and C to be determined graphically. The specific surface area is calculated from these values. This technique assumes that a close-packed monolayer of adsorbed nitrogen is formed, and the cross-sectional area of adsorbed molecules is the same as the closest packing in a solidified gas.

## Density

The density of a powder is the mass of a powder divided by the bulk volume of the powder. Various definitions and measurements of density are used. True density is the density of the actual solid material, exclusive of voids and intraparticle pores larger than the molecular or atomic dimensions in the crystal lattice. True density is determined by using liquid displacement or helium pycnometry. Granule density is exclusive of pores smaller than 10 μm and is determined by using mercury displacement. Bulk density includes voids and intraparticle pores (Martin 1993e). It is a function of true density and packing arrangement, and is also referred to as apparent density.

The packing of particles may be described in terms of void space (Dallavalle 1948e). The porosity ($\varepsilon$) is the fraction of total void volume and is determined from the relationship

$$\varepsilon = 1 - \frac{\rho_b}{\rho} \tag{5.12}$$

where $\rho_b$ is the bulk density and $\rho$ is the true density. In the "rhombohedral" arrangement, which is the closest packing for uniform-sized spheres, each sphere contacts 12 other spheres and has a porosity of 25.95 percent (Dallavalle 1984a). In the "cubical" arrangement, which is the most open packing, each sphere contacts 6 other spheres. This arrangement has a porosity of 47.64 percent. Intermediate arrangements include "orthorhombic" and "tetragonalspheriodal." Real particles have porosity values between 30 and 50 percent.

Bulk density is affected by the particle size distribution, shape, and adhesion properties. Tapping or vibration causes rearrangement of the particles, reducing void space and increasing bulk density. In powders with wide size distributions, smaller particles may fill the void spaces between the larger particles. As particles deviate from sphericity, porosity increases. Increased void space is observed with rough particle surfaces. Smaller particles produce increased porosity due to cohesion. When particles are packed into a cylinder, the voids along the wall are greater than in the body of the bed. In cases where the particle diameter is very small compared with the wall diameter, these wall effects become appreciable.

The bulk density of a powder becomes important in pharmaceutical processes that subdivide the powder by volume. For example, variations in bulk density result in weight and dose variation of tablets.

## Particle Adhesion

Two opposing forces are involved in particle interactions: attraction and detachment. Adhesion is the attractive interaction between two unlike particles, and cohesion is the attractive interaction of similar particles. Detachment forces consist mainly of gravitational and accelerational forces. Both detachment forces depend on mass; thus, the size of the adherent particles will influence the degree of particle interaction.

The many types of adhesion forces may be classified as electrical and nonelectrical. The total force of adhesion at any time is equal to the sum of the individual forces (Stewart 1986).

### Electrical Forces

Contact potential forces and Coulombic forces are the electrical interactions existing in air; double-layer interactions occur in aqueous liquids.

#### *Contact Potential Forces*

Contact potential forces occur during contact between uncharged and unlike particles by electron transfer due to differences in the energy state of the Fermi levels, the outermost conduction band of electrons. Electron transfer occurs on contact between particles until equilibrium is achieved, which results in a contact potential difference. Surface charges may be measured following detachment of the particle. The magnitude of the contact potential force ($F_e$) may be calculated by (Zimon 1982a):

$$F_e = \frac{2\pi q^2}{S} \tag{5.13}$$

where q is the particle charge on detachment from a substrate and S is the contact area between the particle and the substrate. Contact potential forces depend on the sizes and contact areas of the particles and the electron donor-acceptor properties of the materials. The acquisition of electrostatic charges on particles from contact with other particles or surfaces is generally referred to as triboelectric charging. It may occur during such processes as sieving, mixing, milling, and pneumatic conveying.

### *Coulombic Forces*

Coulombic forces occur between charged particles and an uncharged surface. The charged particle induces an equal and opposite charge on the surface, producing image forces. The Coulombic interaction is situated on both sides of the surface at equal distances from the surface. The adhesion force results from the image interaction between particle charges and the induced charge in the substrate. The adhesion force resulting from Coulombic interactions ($F_{im}$) may be determined by (Zimon 1982a):

$$F_{im} = \frac{Q^2}{l^2} \tag{5.14}$$

where Q is the particle charge and l is the distance between centers of the charges. As a result of charge leakage, the magnitude of Coulombic interactions is reduced over time. The rate of charge decay may be exponential or bi-exponential, and depends on relative humidity, temperature, and the presence of adsorbed layers.

### *Electrostatic Double Layer*

Particles in liquid media develop surface charges by selective adsorption of ionic species in the liquid; by ionization of surface groups, such as carboxylate and amine groups; or as a result of differences in the dielectric potential constant between the particle and the liquid (Martin 1993c). The surface charge is balanced by counter-ions of opposite sign, thereby forming a double-layer that surrounds the particle. The closest layer to the particle surface consists of tightly bound solvent molecules and is called the Stern layer. The outer layer (diffuse layer) contains partially mobile ions due to greater solvent mobility. The boundary between the two layers is called the shear plane.

The Derjaguin-Landau-Verwey-Overbeek (DLVO) theory describes the interaction between charged particles in liquids (Visser 1995). As similarly charged particles approach each other, the electrostatic repulsion must be overcome before the separation distance is sufficiently close to allow particle attraction due to van der Waals forces. Increased adhesion is observed with oppositely charged particles.

The zeta-potential is the difference in potential between the shear plane and the electroneutral region of bulk liquid (Martin 1993c). It is affected by the ionic concentration, pH, viscosity, and dielectric constant of the solution and has practical application in the stability of liquid dispersion systems.

### *Electrostatic Charge Measurement*

Electrostatic charge measurement in air is performed by using the Faraday well / cage or electrical low-pressure impactor. Electrostatic charge measurement in liquids is performed by electrophoresis.

**Faraday Well/Cage.** The electrostatic charge of particles is measured by an electrometer after pouring bulk powder into a Faraday well (Staniforth and Rees 1982b) or following particle removal from a Faraday cage in a moving airstream (Kulvanich and Stewart 1987d). The electrostatic charge of particles is expressed as a specific charge for each gram of powder. The electrostatic charge polarity and magnitude was determined from various pharmaceutical excipients following triboelectrification in a cyclone (Staniforth and Rees 1982b) and from detached drug particles in model interactive systems (Kulvanich and Stewart 1987d). Modification of the Faraday well has enabled determination of the electrostatic charge of aerosolized particles (Byron, Peart, and Staniforth 1997).

**Electrical Low-Pressure Impactor.** Electrical low-pressure impactors (ELPIs) allow the aerodynamic size classification and electric charge measurement of aerosol particles in the 0.03 to 10 μm range (Keskinan, Pietarinen, and Lehtimaki 1992). The electrical charge of particles is determined as they impact onto collection stages by using a multichannel electrometer.

**Electrophoresis.** The zeta-potential of particles in liquid media is determined by electrophoresis, which measures the rate of particle movement in an electric field. In an applied electric field, charged particles tend to move toward the oppositely charged electrode. Particles with greater charge move faster. The particle velocity may be characterized by the electrophoretic mobility, which is the ratio of the velocity of the particles to the field strength. The zeta-potential may be calculated from the particle mobility by using the Smoluchowski equation:

$$\zeta = \frac{\eta\mu}{\varepsilon} \tag{5.15}$$

where $\zeta$ is the zeta-potential, $\eta$ is the viscosity of the medium, $\mu$ is the particle electrophoretic mobility, and $\varepsilon$ is the dielectric constant.

The speed of particle movement during zeta-potential measurements was originally determined manually with a microscope. Newer methods use laser Doppler velocimetry (LDV), in which the Doppler frequency shift provides a measure of particle speed.

The stability of colloidal dispersions depends on the zeta-potential of particles. Formulation of stable dispersions should aim at absolute values of zeta-potential above 50 mV, which provide mutual electrostatic repulsion of particles (Rhodes 1989). Coagulation and sedimentation occur when the zeta-potential is close to zero.

## Nonelectrical Forces

Nonelectrical interactions comprise intermolecular forces, capillary forces, and forces due to solid bridge formation.

### *Intermolecular Forces*

Intermolecular forces include interactions predominantly due to van der Waals forces, such as dipole–dipole, dipole–induced dipole, and induced dipole–induced dipole interactions. Other types of intermolecular forces include hydrogen bonding, which is the interaction between a highly electronegative element, such as oxygen in a hydroxyl group of one molecule, and hydrogen from another hydroxyl group.

Van der Waals forces are present in gaseous and liquid environments. Particles may be brought into contact by electrostatic forces, but once adhesion has been established, van der Waals forces keep particles in contact and other forces are required for their detachment (Visser 1995). Electrostatic and capillary forces are smaller in magnitude than van der Waals forces. The van der Waals forces ($F_{vdW}$) are described by (Visser 1989, 1995):

$$F_{vdW} = \frac{Ar}{6z^2} \text{ for a sphere and a plane surface} \tag{5.16}$$

$$F_{vdW} = \frac{Ar}{12z^2} \text{ for two identical spheres} \tag{5.17}$$

where A is the Hamaker constant, r is the particle radius, and z is the distance between the sphere and surface. The Hamaker constant depends upon the molecular properties of the materials involved. Van der Waals forces decrease rapidly with separation distance and are effective only at separation distances smaller than 10 Å (Corn 1961a). The magnitude of van der Waals forces becomes negligible compared with that of gravitational forces when the particle size exceeds 10 μm (Figure 5.1) (Visser 1989).

### *Capillary Forces*

Capillary forces occur following condensation of water vapor between particles, which causes liquid bridge formation (Morgan 1961). The resultant surface tension produces attractive forces. The capillary force ($F_c$) between a spherical particle and a planar surface may be described as (Zimon 1982a; Israelachvili 1991a):

$$F_c = 4\pi r\gamma\cos\theta \tag{5.18}$$

where r is the particle radius, γ is the surface tension, and θ is the contact angle between the particle and the liquid.

**Figure 5.1** Influence of particle diameter on the magnitude of van der Waals forces.

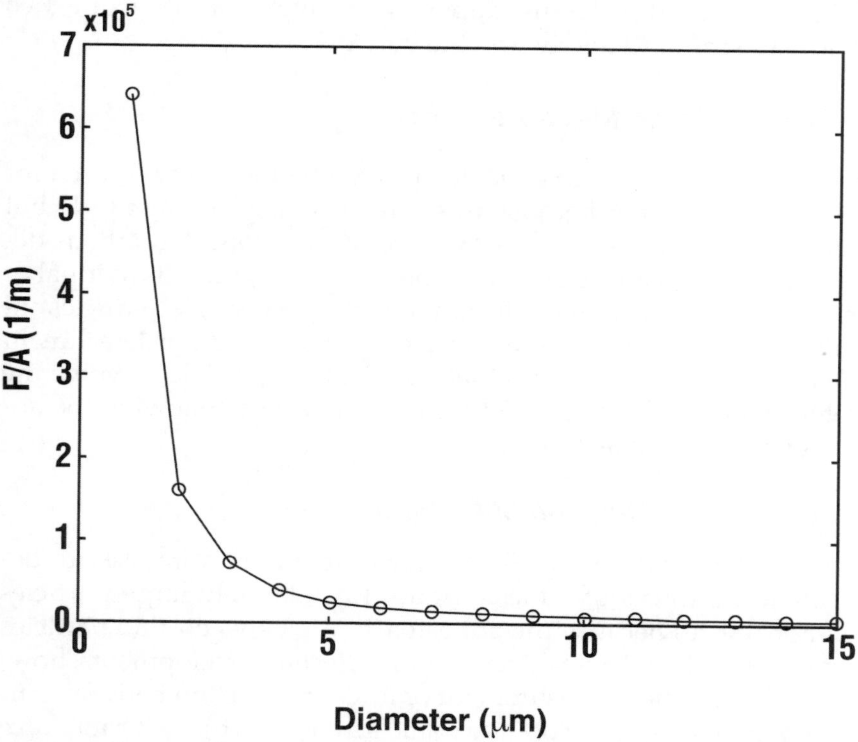


Liquid bridge formation occurs over time; thus, capillary forces are not exhibited immediately after contact of particles to substrates. Factors affecting the magnitude of capillary interactions include relative humidity (RH), particle size, and surface roughness. Generally, in conditions below 50 percent RH, capillary forces do not produce significant adhesion; however, capillary forces will be the dominant adhesion force between 65 percent and 100 percent RH (Zimon 1982a).

### *Solid Bridges*

Solid bridges between particles may be formed by chemical reaction, partial melting followed by subsequent solidification, or partial dissolution followed by subsequent crystallization (Stewart 1986; Morgan 1961). Solid bridge formation occurs over time and reinforces other types of preceding interactions. The formation of solid bridges following drying depends on hygroscopic properties (Padmadisastra, Kennedy, and Stewart 1994a).

## Adhesion Force Measurements

The force of adhesion may be determined by the force required for particle detachment because this force is equal in magnitude but opposite in sign to the adhesion force (Zimon 1982e). Methods developed for measuring adhesion forces involve either individual or bulk particle detachment. Techniques that use single particles involve the direct measurement of the force required for detachment of each particle. Techniques that use multiple particles involve the determination of a number of particles detached following the application of a specific force.

### *Single-Particle Detachment Methods*

Historically, single-particle detachment methods were used to determine the forces of adhesion only for relatively large particles (Zimon 1982e) because the adhesion forces between fine particles were too small to detect (Morgan 1961). Recent developments, however, have enabled measurement of forces interacting between surfaces at much higher resolution (Israelachvili 1991b). Adhesion force may be measured between two spheres or a sphere and a plane surface. Methods for determining adhesion force by using single-particle detachment include microbalance, pendulum, and atomic force microscopy.

**Microbalance Method.** The microbalance method measures the adhesion force of a spherical particle attached to the end of a quartz fiber, which acts as a cantilever, helical spring, or torsion balance (Corn 1966). The force of adhesion is measured by the elongation of the quartz spring as follows (Zimon 1982e):

$$F = c\Delta h \tag{5.19}$$

where c is a parameter characterizing spring rigidity and $\Delta h$ is the elongation of the spring at the moment of detachment (Figure 5.2).

**Figure 5.2** Methods for measuring particle adhesion.

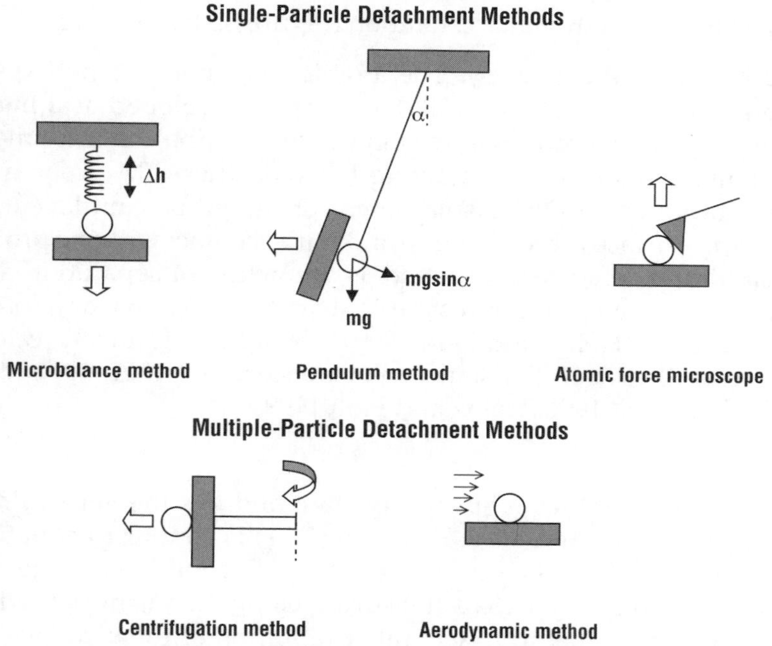


The preliminary compression force producing particle adhesion affects the force of interaction due to differences in true contact area between particles (Zimon 1982e). The microbalance method was used to determine the adhesion force of simulated quartz and Pyrex™ particles to various surfaces, where increased particle size or humidity increases adhesion, and increased surface roughness reduces adhesion (Corn 1961b).

**Pendulum Method.** The pendulum method determines the adhesion force by calculating the detachment angle of a freely hanging sphere acting under gravitational force. Detachment occurs in a perpendicular direction to the contact area, and the adhesion force is determined from the angle of detachment as follows (Zimon 1982e):

$$F = mg \sin \alpha \tag{5.20}$$

where $\alpha$ is the angle of deviation from the vertical, m is the mass, and g is the gravitational acceleration (Figure 5.2).

**Atomic Force Microscopy.** The atomic force microscope (AFM) (Binnig, Quate, and Gerber 1986), originally developed as a high-resolution imaging tool, is now being used for adhesion force measurements. Adhesion forces acting between the AFM probe and sample produce vertical displacement of the probe cantilever. As the sample surface is brought into and out of contact with the probe, the cantilever deflection is plotted as a function of separation distance between the probe and sample. This plot is known as a force-distance curve (Heinz and Hoh 1999). The adhesion force is calculated from these force-distance plots, according to Hooke's law (Shakesheff et al. 1996; Heinz and Hoh 1999):

$$F = kx \tag{5.21}$$

where k is the cantilever spring constant and x is the vertical displacement of the cantilever. Equation (5.21) is similar to equation (5.19) for the microbalance. The vertical deflection is measured in the AFM technique, whereas the spring elongation is measured in the microbalance method. Accurate calibration of the cantilever is required to provide quantitative force measurements.

Use of the colloidal probe technique allows measurement of the adhesion force between a spherical colloid particle attached to the tip of the AFM cantilever and the sample surface (Ducker, Senden, and Pashley 1991) (Figure 5.2). The colloid probe technique has been used to measure adhesion forces between particles and surfaces in air (Shakesheff et al. 1996) as well as in liquids (Ducker, Senden, and Pashley 1991). The use of a spherical colloidal particle provides a well-defined geometry for sample contact (Claesson et al. 1996). Particle adhesion is dependent on the surface roughness of either the particle or the surface and the contact geometry between the probe and sample. Increased adhesion may occur when a particle fits snugly in pits or grooves of the surface, whereas decreased adhesion may occur when particles sit atop ridges or bumps on the surface (Mizes 1995). A higher applied load may result in higher adhesion due to increased contact area.

AFM studies have been used to investigate the adhesion force between spheres or particles. As examples, Fuji et al. (1999) observed increased adhesion between two silicon spheres in humid conditions (70 percent RH) due to capillary condensation. Sindel and Zimmerman (1998) measured the adhesion forces between two lactose particles, and a lactose particle and compressed lactose and sapphire substrates in a vacuum and air. Louey, Mulvaney, and Stewart (2001) determined the adhesional properties of lactose carriers for dry powder inhalers by using a colloid probe. The adhesion forces at individual sites were stable, and the lognormal force distributions obtained enabled characterization by the geometric mean force and geometric standard deviation. The sensitivity of the technique was sufficient to differentiate between the samples examined. The ability to characterize surfaces of drug delivery systems using AFM has the potential to advance the understanding of the relationship between surface properties and delivery system functionality (Shakesheff et al. 1996).

### *Multiple-Particle Methods*

Methods for measurement of bulk-particle detachment forces include the use of centrifugation, vibration, aerodynamic force, and impaction. For all of these techniques, the detaching force acts simultaneously on all of the adherent particles present on the surface.

**Centrifugation.** The centrifuge method determines adhesion following particle detachment from the adhering surface during rotation around an axis (Figure 5.2). The magnitude of the detaching force (F) can be determined by the following relation (Zimon 1982e):

$$F = V(\rho - \rho_0)(j + g) \tag{5.22}$$

with

$$j = \omega^2 l \tag{5.23}$$

and

$$\omega = \frac{2\pi n}{60} \tag{5.24}$$

where V is the particle volume; $\rho$ and $\rho_0$ are the density of the particle and surrounding media, respectively; j is the centrifugal acceleration; g is the gravitational acceleration, $\omega$ is the angular rotating velocity of the surface, l is the distance between the surface and the axis of rotation; and n is the number of revolutions of the surface per minute. For adhesion in air, where $\rho >> \rho_0$ and $j >> g$, the force of detachment may be simplified to (Zimon 1982e):

$$F = V\rho\omega^2 l \tag{5.25}$$

The adhesion force distribution is obtained by measuring the number or mass of detached particles at various centrifuge speeds. It generally follows a lognormal relationship, enabling characterization by the geometric mean force ($F_{50}$) and the geometric standard deviation (GSD) (Zimon 1982e) or by the mean centrifuge speed ($S_{50}$; the speed required to detach 50 percent of adherent particles) and GSD (Kulvanich and Stewart 1987c, 1987e). Particle detachment may be determined optically (Kordecki and Orr 1960; Boehme et al. 1962; Podczeck, Newton, and James 1994) or by chemical analysis (Kulvanich and Stewart 1987e; Staniforth et al. 1981). The centrifuge speed must be increased or decreased smoothly to eliminate the influence of inertial forces, and each speed should be held for several seconds. However, increasing the centrifuging time does not have any effect on particle detachment in air.

The centrifuge technique has been used to determine adhesion forces between pharmaceutical particles and substrate surfaces (Booth and Newton 1987; Lam and Newton 1991) or compressed powder surfaces (Podczeck, Newton, and James 1994, 1995b; Okada, Matsuda, and Yoshinobu 1969). Increased centrifugal forces obtained by the ultracentrifuge enabled determination of the adhesion force between micronized particles and compressed powder surfaces (Podczeck, Newton, and James 1995a). Detached drug particles were separated from intact interactive units by using a wire mesh (Staniforth et al. 1981, 1982; Kulvanich and Stewart 1987c, 1987e) or a brass plate holder (Laycock and Staniforth 1983). However, distortion of the mesh occurred at high rotation speeds; in addition, disintegration and flattening of the carrier were encountered (Laycock and Staniforth 1984). The detachment angle was not always 90°, as expected for the ideal sphere resting on a planar surface. Particles located in clefts on the carrier surface were proposed to roll or slide along a surface and dislodge other particles in their path before detachment (Laycock and Staniforth 1984). Particle-on-particle adhesion was examined by attachment of interactive mixtures to an aluminum support disk (Podczeck, Newton, and James 1995c).

A modified centrifuge technique determined the frictional forces between particles and surfaces by mounting the surface at angles between 0° and 90° (Podczeck and Newton 1995). The force of detachment ($F_{det}$) decreased with decreasing angle of detachment ($\alpha$). A linear relationship was obtained by plotting $F_{det}$ cos $\alpha$ against $F_{det}$ sin $\alpha$. The coefficient of friction ($\mu$) and the frictional force ($F_{frict}$) were determined from the slope and intercept, respectively.

**Vibration.** The magnitude of the detaching force (F) by vibration can be determined by the following relation (Zimon 1982e):

$$F = m(j + g) \tag{5.26}$$

with

$$j = 4\pi^2 v^2 y \cos\left(wt + \frac{\pi}{2}\right) \tag{5.27}$$

and

$$w = 2\pi v \tag{5.28}$$

where m is the particle mass; j is the vibrational acceleration; g is the gravitational acceleration; v is the vibration frequency; y is the vibration amplitude; and t is time. The detachment force is adjusted by varying the vibration frequency and amplitude. More particles are detached by the vibration method at relatively small forces of detachment than by the centrifuge method. However, centrifugation detaches more strongly adhered particles than does vibration. Particle deformation may occur with the vibrational force; thus substrates subject to plastic deformation are not suitable for this technique (Zimon 1982e).

The vibration technique has been employed to determine adhesion force and segregation tendency of powder mixtures (see, for example, Yip and Hersey 1977; Bryan, Rungvejhavuttivittaya, and Stewart 1979; Stewart 1981; Staniforth and Rees 1982a, 1983; Staniforth, Ahmed, and Lockwood 1989). The vibration apparatus consists of either stacking cylinders or a single sieve. The dislodged drug particles pass through the sieve aperture, whereas the intact interactive unit is retained above the sieve. However, comminution of carrier particles enables undetached drug particles to pass through the sieve (Stewart 1981).

**Aerodynamic Force.** An adherent particle will be dislodged by an airstream flowing across the surface (Figure 5.2) when the aerodynamic force provided by the drag ($F_{dr}$) meets or exceeds the force of adhesion ($F_{ad}$), particle weight (P), and lift of the airstream on the particle ($F_{lif}$) (Zimon 1982c):

$$F_{dr} \geq \mu(F_{ad} + P + F_{lif}) \tag{5.29}$$

where μ is the coefficient of friction.

The aerodynamic force of drag may be described by the relation (Zimon 1982c):

$$F_{dr} = \frac{c_x \rho v^2 \pi r^2}{2} \tag{5.30}$$

where $c_x$ is the drag coefficient; ρ is the air density; v is the airflow velocity; and r is the particle radius. The adhesion force distribution is obtained by measuring particle detachment at various airflow velocities. The distribution follows a lognormal distribution, enabling characterization by the median detachment velocity and standard deviation (Zimon 1982c). This technique is also referred to as the hydrodynamic method (Mullins et al. 1992). The detachment velocity is dependent on the particle size and shape and the surface roughness.

**Impaction.** The impact technique determines adhesion force from the detachment of particles by the impact of a hammer or bullet from the opposite face of the adhering surface. The magnitude of the detaching force (F) can be determined by the simple force relation (Otsuka et al. 1983; Concessio, Van Oort, and Hickey 1998):

$$F = ma \tag{5.31}$$

where m is the impacting mass and a is the impact acceleration.

A pendulum impact technique using a swinging hammer was employed to determine the adhesion force between organic and pharmaceutical particles and a substrate surface (Otsuka et al. 1983; Concessio, van Oort, and Hickey 1998). The impaction force was adjusted by changing the lift angle of the impact hammer and the hammer weights. The number of particles detached after impaction was determined. A lognormal force distribution was obtained, enabling characterization by $F_{50\%}$, the separation force at which 50 percent of the particles remained on the substrate (Otsuka et al. 1983). The effect of particle shape, surface roughness, and the presence of fine particles on the adhesion force was examined by this technique (Otsuka et al. 1988). The median particle size of excipients that remained adhered to the substrate surface was correlated to powder flow characteristics (Concessio, Van Oort, and Hickey 1998; Concessio et al. 1999).

## Factors Affecting Particle Adhesion

Particle adhesion is affected by fundamental particle properties, such as particle size and shape and surface roughness, and environmental factors, such as storage duration and humidity.

Increasing the particle size causes an increase in adhesion force (Corn 1961b). However, the forces of detachment (acceleration and gravitational forces) are dependent on particle mass. Thus, smaller particles have relatively greater forces of adhesion as compared with forces of detachment due to their reduced particle mass (Corn 1966). Gravitational forces become negligible at particle diameters less than 10 μm, and adhesion forces become more dominant (Visser 1995). Particle detachment is proportional to particle diameter (d) for the microbalance method, whereas detachment force is proportional to $d^3$ for the centrifugation and vibration methods and to $d^2$ for the aerodynamic method (Corn 1966). Greater particle detachment with increased particle size has been observed when using the centrifugation (Zimon 1982d; Kulvanich and Stewart 1987c) and vibration techniques (Nyström and Malmqvist 1980).

Theoretical equations for adhesion assume that particles are ideally smooth and spherical. Particle shape and surface roughness may affect particle adhesion due to their effect on contact area and separation distance. Surface roughness may increase the separation distance between particles, thus reducing van der Waals forces, which become negligible at approximately 10 Å (Corn 1961a). Surface roughness may be described as ideally smooth, or having microroughness and macroroughness (Figure 5.3) (Zimon and Volkova 1965). Compared to ideally smooth surfaces, reduced contact area and adhesion occurs when particles contact the top of the asperities on surfaces with microroughness. Conversely, increased contact area and adhesion occurs when particles contact the valley between two asperities on surfaces with macroroughness. The effect of surface roughness depends on the size of the asperity, the interval distance between asperity peaks, and the relative size of the adhering particle (Zimon 1982b). Increased physical stability of interactive mixtures containing macrorough carrier particles was found to be due to multiple adhesion sites and entrapment of the adherent particles in surface clefts, which prevents detachment caused by rolling, abrasion, or collision with other particles during further mixing or processing (Staniforth 1987). Indeed, surface roughness may increase interparticulate forces by mechanical interlocking (Staniforth 1988). Increased adhesion of salicylic acid particles on Emdex™, Dipac™, and lactose carriers was attributed to greater surface roughness (Staniforth et al. 1981).

The orientation of nonspherical particles adhered to a surface may alter the contact area. Plate-shaped particles may have increased adhesion due to their flat contact area. Increased adhesion was observed as particles became more spherical, an effect attributed to increased contact area (Otsuka et al. 1988). In contrast, the reduced adhesion observed in particles with surface protuberances or adhered fine particles was attributed to reduced contact area (Otsuka et al. 1988).

**Figure 5.3** Types of surface roughness affecting particle adhesion: (A) ideally smooth, (B) microroughness, and (C) macroroughness. Adapted from Zimon and Volkova (1965).

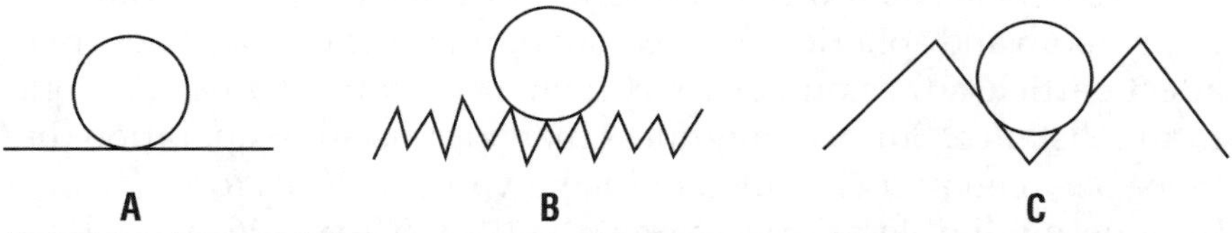

The chemical composition of particles affects their deformity, strength, and electrical properties (Neumann 1967). The chemical and electrical nature of particles affects their adhesion due to the influence of these factors on van der Waals, electrostatic, and capillary forces (Corn 1966). In addition, the viscoelastic properties of the particles influence the degree of surface deformation and contact area of adhesion (Corn 1966). Plastic deformation causes permanent deformation and increases adhesion. Conversely, elastic deformation produces reversible deformation and may indirectly affect adhesion due to its influence during particle detachment (Visser 1995). Contact areas increase due to deformation as time of contact is extended (Visser 1995). Particle adhesion with elastic deformation may be described by using the Derjaguin-Muller-Toporov (DMT) or Johnson-Kendall-Roberts (JKR) models. The DMT model is applicable for hard materials with Young's modulus greater than $10^9$ N/m$^2$, and considers the interaction as a ring-shaped zone around the contact area. The JKR model is more appropriate for soft materials and assumes the interaction force exists at the contact area only (Visser 1995; Ziskind, Fichman, and Gulfinger 1995).

Differences in polymorphic state and degree of crystallinity may affect adhesion due to differences in physicochemical properties. The formation of capillary forces depends upon the moisture adsorption capabilities of the contacting materials and the time required for liquid bridge formation. Increased adhesion after storage in humid conditions was observed in mixtures that contained carriers with high moisture affinity, due to capillary interactions, whereas no change was observed in carriers with minimum moisture sorption (Padmadisastra, Kennedy, and Stewart 1994a). Crystallization of amorphous regions may result in solid bridging and permanent agglomeration of particles (Ward and Schultz 1995).

Electrostatic charge influences particle adhesion directly through electrostatic interactions and indirectly by increasing van der Waals interactions due to reduced separation distance. Oppositely charged drug and carrier particles have been found to facilitate close surface contact and increase the strength of van der Waals forces (Staniforth 1987). The stability of binary and ternary interactive mixtures was affected by the electrostatic charges of the adhering particles (Staniforth et al. 1982).

Environmental factors, such as mixing and storage duration and conditions, may affect the adhesional properties of powders. Increased adhesion occurred in interactive mixtures with increased blending times (Kulvanich and Stewart 1987b). Possible explanations for this include increased triboelectric charging, the presence of drug aggregates at shorter blending times enhancing drug detachment, greater intermolecular interaction due to increased impact and surface deformation with time, or fracture of particles with increased blending time enabling adhesion of new surfaces.

Applied force during particle contact may affect particle adhesion. Increasing the applied force increased particle adhesion by larger contact area caused by plastic deformation (Podczeck, Newton, and James 1994; Booth and Newton 1987). Plastic deformation was material-dependent and was related to yield pressure (Lam and Newton 1991). Increased adhesion occurred with increased duration of applied pressure, due to the time dependence of plastic deformation (Lam and Newton 1993). An increase in temperature increased the plasticity of materials, thus enhancing adhesion (Lam and Newton 1992a). Larger particles produced increased contact area and adhesion with plastic deformation under applied force (Lam and Newton 1992b). Greater increases in adhesion occurred with increasing applied force for irregular-shaped particles compared to spherical particles (Podczeck, Newton, and James 1994).

During storage at low relative humidities, the electrostatic charge reduces due to a process known as "charge decay," thus reducing the total adhesional force (Zimon 1982a; Staniforth and Rees 1982b; Kulvanich and Stewart 1987a). The rate and extent of charge decay was dependent upon the relative humidity and particle system. For example, the rate of charge decay increases with increasing relative humidity (Kulvanich and Stewart 1988). In addition, at increasing relative humidity, van der Waals forces increase because the adsorbed moisture layer produces reduced distances between particles.

Increased adhesion was observed following storage above 65 percent RH because capillary condensation formed a liquid layer (liquid bridge) in the gap between the contacting bodies (Zimon 1963). Increased adhesion due to capillary forces also has been observed following storage at high humidity (Padmadisastra, Kennedy, and Stewart 1994a; Podczeck, Newton, and James 1996, 1997). Solid bridge formation was observed following oven drying (100°C) or column-drying of mixtures stored at high relative humidities (Padmadisastra, Kennedy, and Stewart 1994a, 1994b).

## Particle Motion in Bulk Powders

Particle motion in bulk powders may be observed in such processes as powder flow, fluidization, and mixing. These processes are described in the following sections.

### Powder Flow

Powder flow requires the expansion of the packed bed, individual particles to overcome interparticle forces, and increased particle separation distance, all of which combine to allow particle mobility. The flow of powders may be characterized in four phases—plastic solid, inertial, fluidization, and suspension (Crowder and Hickey 2000). At rest, powders exhibit a plastic solid phase, characterized by small interparticle spacing and slow particle velocities. The powder bed stresses are independent of velocity. When the powder is tilted above its angle of repose, transition to the inertial phase occurs. In the inertial phase, the interparticle spacing increases, particle collisions produce powder bed stresses, and powder flow begins but is confined to the surface layer. The inertial phase is the powder flow characterized in pharmaceutical powders. When the interstitial fluid provides sufficient momentum to overcome interparticle forces, fluidization occurs. During fluidization, the interparticle spacing is of the order of the particle size. Very large interparticle spacing, negligible interparticle forces, and particle velocities similar to the entraining fluid velocity characterize the suspension phase.

Factors affecting powder flow are related to those affecting particle adhesion and cohesion (Staniforth 1988). These include particle size, shape, porosity, density, and surface texture (Martin 1993e). Powder flow is reduced as particle size decreases, due to the greater adhesion forces relative to gravitational forces. Generally, particles greater than 250 μm are free-flowing, particles smaller than 100 μm have reduced flow, and particles smaller than 10 μm are very cohesive and have poor flow under gravity, except as large agglomerates (Staniforth 1988). The presence of particles smaller than 10 μm promotes cohesiveness and may reduce powder flow (Neumann 1967).

The flow properties of such powders may be improved by removing the "fines" or adsorbing them onto added larger particles (Martin 1993e). Particle shape influences powder flow due to the contact area between particles. Optimal flow properties are provided by spherical particles, which have minimal interparticulate contact. Irregular-shaped particles, such as flakes or dendrites, have high surface-to-volume ratios and poorer flow properties (Staniforth 1988). Particle density affects powder flow through its influence on the relative contributions of gravity and surface forces. Denser particles generally have better flow properties (Neumann 1967). Surface roughness affects powder flow due to its influence on particle adhesion. Rough particles have greater tendency for mechanical interlocking than particles with smooth surfaces (Staniforth 1988). Moisture adsorption on the particle surface may improve powder flow by the combined effects of pore-filling, increased particle density, and lubrication to a certain level, whereas capillary forces produce reduced flow (Neumann 1967).

Methods for measuring powder flow are based on static bed or dynamic flow properties, as discussed next.

## Static Methods

Static bed methods include angle of repose, angle of spatula, Carr's compressibility index, Hausner ratio, and shear cell measurements.

### *Angle of Repose and Angle of Spatula*

When a bulk powder is allowed to fall freely and accumulate on a flat surface, it forms a cone of material. The angle of repose—the angle that the surface of the cone makes with the horizontal surface—has been stated to be an indirect measure of particle size, shape, porosity, cohesion, fluidity, surface area, and bulk (Carr 1965). It provides an indication of the ease at which the powder will flow. Small angles of repose (less than 40°) indicate free-flowing powder, whereas large angles (greater than 50°) indicate poor flow properties. However, the angle of repose does not provide any indication of the flow pattern or flow rate. The angle of repose increases as the particle departs from sphericity and as the bulk density increases. It is independent of particle size when the particle diameter is larger than

100 μm, but it increases sharply with particle diameters smaller than 100 μm. The value of the angle of repose can differ, depending on which of the various measurement methods are used.

The angle of spatula is similar to the angle of repose. It is the angle of powder remaining on a spatula that has been vertically removed from a powder bed. An average angle is measured prior to and after gentle tapping. For a given material, the angle of spatula is generally higher than the angle of repose (Carr 1965). Similar to the angle of repose, smaller values are indicative of better flow.

### *Carr's Compressibility Index and the Hausner Ratio*

Powder flow properties may be empirically estimated by using the porosity and packing density of powders. Carr's compressibility index and the Hausner ratio are examples of these estimates.

Carr's compressibility index empirically estimates the powder flow from packing density and compressibility. Carr's compressibility index (CI) is based on the difference between the tapped and bulk density, and is expressed as a percentage:

$$CI = \frac{\rho_{tap} - \rho_{bulk}}{\rho_{tap}} \cdot 100\% \tag{5.32}$$

where $\rho_{tap}$ is the tapped density and $\rho_{bulk}$ is the bulk density of a powder. Compressible powders are expected to be more cohesive and exhibit poor flow; thus, powders with lower CI values are estimated to have better flow behavior. Powders with CI values greater than 20 percent have been found to exhibit poor flow properties (Carr 1965).

The Hausner ratio estimates powder flow based on packing density and compressibility, similar to Carr's compressibility index. The Hausner ratio (HR) is calculated by:

$$HR = \frac{\rho_{tap}}{\rho_{bulk}} \tag{5.33}$$

### *Jenike Shear Cell*

A shear cell apparatus was developed to characterize powder flow by measuring the shear stress at different values of normal stress

(Jenike 1954). The powder is placed under a normal load until plastic consolidation occurs. A horizontal shear force is applied at reduced consolidation loads, and the force required for shear movement is measured (Amidon 1995; Smith and Lohnes 1984). Various parameters are obtained from the yield locus, which is defined as the relationship between the shear strength of a material and the applied load. The equation for the yield locus is (Amidon 1995):

$$\frac{\tau}{\tau_p} = \left( \frac{\sigma + S}{\sigma_p + S} \right)^{\frac{1}{n}} \tag{5.34}$$

where $\tau_p$ is the shear strength of the powder bed under the consolidation load $\sigma_p$; $\tau$ is the shear strength of the powder bed under a reduced load $\sigma$; S is the tensile strength of the powder; and n is the shear index. The tensile strength (S) is the negative load at which the shear strength is zero, and the shear index (n) is a measure of the degree of curvature of the yield loci. Values of n lie between 1 and 2; n approaches 1 for free-flowing powders and approaches 2 for poor-flowing powders. Other parameters obtained from the yield locus include the angle of internal friction ($\phi$), which is a measure of the shear strength of material per unit of consolidation force, and the unconfined yield stress ($f_c$), which is a measure of the stress required to cause a failure in shear when the material is unsupported in two directions. The angle of internal friction is equal to the angle of repose.

## Dynamic Methods

Dynamic flow methods include the rotating powder drum and the vibrating spatula.

### *Rotating Drum*

As a powder sample is slowly rotated in a drum, the powder rises until its angle of repose is exceeded and an avalanche occurs. The avalanching behavior is characterized by either the time interval between avalanches or the angle of repose. Although an identical rank-order correlation of flow properties was determined in a study using time-to-avalanche data, better differentiation between powders was obtained by using the angle of repose data (Crowder et al. 1999).

The dynamic angle of repose oscillates around a mean value. The change of angle of repose with respect to the mean angle plotted against time produces an oscillating data plot. The change of angle of repose with respect to the mean angle plotted against the mean angle of repose produces a phase space attractor plot (Hickey and Concessio 1997), where data points cluster around a central attractor point. The scatter around this attractor point represents a measure of variability in the avalanching behavior. Free-flowing powders have a tight attractor, whereas cohesive powders have widely fluctuating attractor plots. Lower attractor points indicate better flowability (Kaye 1997). Oscillating data and phase-space attractor plots may also be used as indicators of deterministic chaos, by which periodic, random, and chaotic patterns can distinguish different powder behaviors. Mathematical analysis is used to derive fractal dimensions and Hausdorff, or capacity dimensions (Hickey and Concessio 1997).

### *Vibrating Spatula*

The vibrating spatula characterizes dynamic powder flow by use of the fractal dimension (Hickey and Concessio 1994; Crowder and Hickey 1999). The mass of powder flowing from a vibrating spatula is measured as a function of time. Fractal analysis is performed on the mass versus time profile, and the fractal dimension is determined. Generally, fractal dimension values range between 1 and 2, with lower values indicative of more regular flow behavior. This method has been used for characterization of lactose, sodium chloride, and maltodextrin particles (Hickey and Concessio 1994; Crowder and Hickey 1999).

## Fluidization

As a fluid (gas or liquid) flows upward through a particle bed at a low flow rate, the fluid percolates through the interstices of the fixed bed. As the flow rate increases, the bed expands as the particles move apart until they become suspended in the fluid. The incipient fluidization point occurs when the drag force exerted by the fluid counterbalances the gravitational force on the particles. In liquid systems, smooth progressive bed expansion occurs as the flow rate increases, and particles are homogeneously distributed thoughout the bed. In gas systems, fluidization depends on the particle properties and interparticulate forces. The capacity of powders to be fluidized by gas depends on the rate at which they expand (Rietma 1991c). Higher gas fluidization rates produce greater expansion and better powder flow, unless particle cohesion hinders the penetrating gas, producing slow dilution and bubbling.

Gas fluidization of powders has been empirically characterized into four groups by using the Geldart classification (Figure 5.4) (Geldart 1986).

Group A powders are slightly cohesive powders. At low gas velocities, they are homogeneously fluidized in a stable expanded

**Figure 5.4** Geldart fluidization classification of powders. Adapted from Geldart (1986).

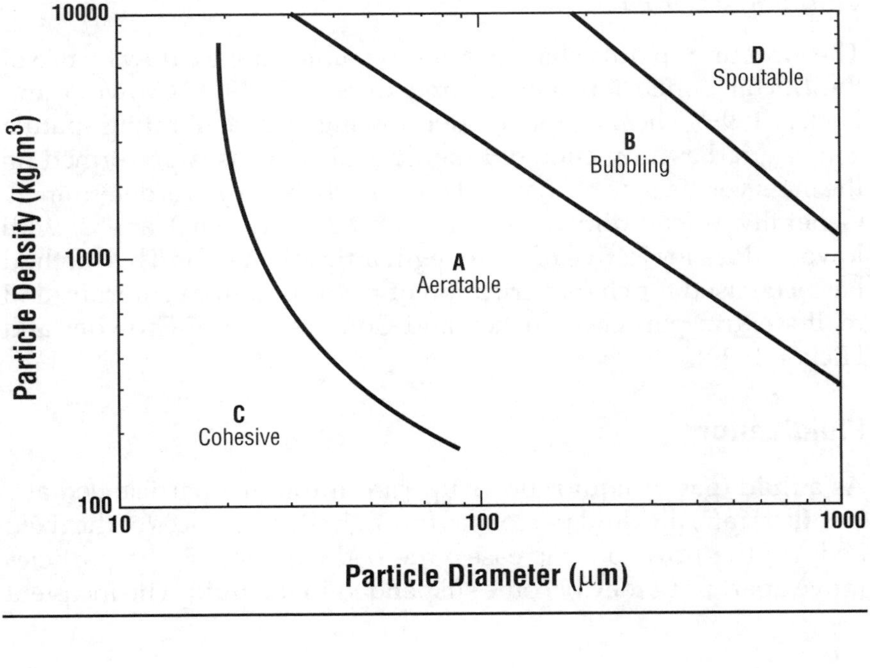

state (Rietma 1991c). At higher gas velocities, heterogeneous fluidization occurs with rapid mixing and bubbling of gas through the powder bed, similar to boiling liquid. The bubble size is affected by particle size, proportion of fine particles (smaller than 45 μm), pressure, and temperature. The powder bed undergoes aeration and large powder bed expansion (Geldart 1986).

Group B powders exhibit negligible interparticle forces. Fluidization occurs with bubbling, slightly above the incipient fluidization. The bubbles burst at the surface as discrete entities. Minimal aeration and powder bed expansion occurs. The powder bed collapses rapidly when the fluidization gas is removed. Little or no powder circulation occurs in the absence of bubbles (Geldart 1986).

Group C powders are cohesive and extremely difficult to fluidize, due to the stronger interparticle forces compared with the fluidizing force. The particles are generally smaller than 20 μm, soft or irregular-shaped (Geldart 1986). Channelling and plugging occurs at velocities greater than the incipient fluidization gas velocity, and the gas escapes through vertical channels in the powder bed while the remainder of the powder bed remains stagnant with very little expansion. This action is attributed to the high cohesion between individual particles (Rietma 1991c).

Group D powders consist of large or dense particles. During fluidization, spouting of the fluidizing gas occurs. The bubbles rise more slowly than the interstitial fluidizing gas, causing the gas to flow into the base of the bubble and out of the top. This results in relatively poor mixing of solids. Size segregation of particles is likely to occur (Geldart 1986).

## Mixing

Powder mixing depends on the particle interactions. Random mixing occurs between noninteracting particles, whereas interactive mixing involves particle adhesion.

### *Random Mixing*

Random mixing occurs with statistical randomization, in which the probability of finding one type of particle at any point in the mixture is equal to the proportion of that type of particle in the mixture. Mixing and demixing processes occur simultaneously. The rate of mixing follows first-order decay. The three principal mechanisms involved in mixing include diffusion, shear, and convection (Lantz and Schwartz 1981). Diffusion produces the redistribution of particles by random movement of the particles relative to one another, analogous to gaseous diffusion. Shear mixing involves the three-dimensional movement of adjacent particles due to formation of slip planes within the powder bed. Convection produces the movement of groups of adjacent particles to another part of the powder bed. Diffusion is referred to as micromixing, whereas shear and convection are referred to as macromixing. Factors affecting random mixing include particle size, shape, density, friability, and mixing proportions. Segregation in random mixtures may occur by percolation and densification, by which smaller particles percolate through void spaces, due to agitation and vibration or gravity, and produce caking at the bottom of the powder bed under the force of the weight above. Segregation may also occur with particle acceleration. Different sized particles have different stopping distances when projected in a horizontal direction, which may result in trajectory segregation.

### *Interactive Mixing*

Interactive mixing generally involves the adhesion of small particles (usually the drug) onto the surface of a larger carrier particle (usually the excipient). Interactive mixtures, originally known as ordered mixtures, produce stable homogeneous mixes due to particle adhesion (Hersey 1975). During mixing, high energy is required to break up the cohesive forces within agglomerates of small particles (Hersey 1979). Factors affecting particle adhesion within interactive mixtures include the size, shape, and surface morphology of both drug and carrier particles, drug concentration, and presence of ternary components. Two types of segregation mechanisms exist for interactive mixtures. Constituent segregation occurs when the adhered drug particles become detached from carrier particles. Interactive unit segregation occurs when interactive units of different sizes undergo percolation or trajectory segregation.

Mixing equipment may be classified as batch or continuous mixers (Lantz and Schwartz 1981). The entire powder sample is blended and removed in batch mixers. The energy input varies greatly between mixers. Low-energy units, such as V-shaped blenders, use rotation of the shell for mixing. High-energy units, such as ribbon blenders, use rotating blades to break up powder aggregates. Energy input is required to overcome the cohesive forces within particle aggregates so the separate particles can move in relation to one another (Carstensen 1980). Low-energy mixers may not be able to provide sufficient energy for the blending of cohesive and electrostatic powders. In these cases, the agglomerates rather than the individual particles are mixed.

## Particle Motion in Gaseous Dispersions

The dynamics of particles in gaseous systems are described in different size regimes: the free molecule regime for particles smaller than 0.01 μm, the transition regime for particles in the 0.01–0.4 μm range, the slip flow regime for particles larger than 0.4 μm, and the continuum or Stokes' regime for particles larger than 1.3 μm (Hesketh 1986a).

### Free Molecule Regime

Particles in the free molecule regime undergo thermal diffusion due to Brownian motion as a result of the small particle size compared to the mean free path of gas molecules. Brownian motion is the constant random movement of small particles suspended in a fluid caused by their constant bombardment by fluid molecules. Each particle moves with uniform velocity, with smaller particles moving more rapidly than larger ones. Particles in high concentrations move more rapidly than those in dilute concentrations. Also, particles move more rapidly through media with lower viscosity. The amplitude of motion is directly proportional to the absolute temperature. The diffusion of the particles depends on the motion of the fluid (Dallavalle 1948d). These particles remain suspended due to the gas convective motion and the high particle diffusion force, which is more significant than the gravitational settling force (Hesketh 1986b).

## Continuum (Stokes') Regime

Particle movement in the continuum size regime follows Stokes' law. A particle falling freely under gravitational forces accelerates until a terminal velocity is attained. The terminal settling velocity (v) in laminar conditions is given by Stokes' law:

$$v = \frac{d^2 g(\rho_s - \rho_0)}{18\eta} \tag{5.35}$$

where d is the particle diameter; g is the gravitational acceleration; $\rho_s$ is the particle density; $\rho_0$ is the density of the dispersion medium; and $\eta$ is the viscosity of the dispersion medium. Stokes' law assumes that particles are rigid spheres, the fluid is incompressible, the fluid velocity at the particle surface is zero, and the particle motion is constant and not affected by other particles or nearby walls.

The nature of motion (laminar, intermediate, or turbulent) depends on the particle size, velocity, and the properties of the surrounding fluid. Reynold's number, Re, is a nondimensional number that describes the nature of motion (Hinds 1982d):

$$Re = \frac{\rho_0 dv}{\mu} \tag{5.36}$$

where $\rho_0$ is the density of the fluid; d is the particle diameter; v is the relative velocity between the particle and fluid; and $\mu$ is the coefficient of viscosity. Re represents the ratio of inertial forces to frictional forces acting on the particle. Laminar flow is observed in spherical particles with Re values smaller than 1. As Re values increase above 1, eddies form on the downstream side of the particle and the flow is turbulent. Particle shape and orientation during settling may profoundly affect the drag force acting on the particle and settling velocity. The nature of flow in pipes is also determined by Re. The boundary of Re values between laminar and turbulent flow differs between flow around particles and within pipes.

The resistance to motion is due to the opposing drag force of the particle. The drag force increases with particle size and deviations from sphericity. The coefficient of drag, $C_D$, is a correctional factor used to account for deviation in Stokes' law due to particle size and shape factor. The value of $C_D$ is dependent on Re (Hesketh 1986b):

$$C_D = \frac{24}{Re} \text{ for Re} < 0.1 \quad (5.37)$$

$$C_D = \frac{24}{Re} + 4.5 \text{ for } 0.1 < \text{Re} < 10 \quad (5.38)$$

and

$$C_D = \frac{24}{Re} + \frac{4}{Re^{\frac{1}{3}}} \text{ for Re} > 10 \quad (5.39)$$

For particles with Re values greater than 1000, $C_D \cong 0.44$, and particle movement follows Newton's law, of which Stokes' law is a special case (Hinds 1982e). The relationship between particle shape and the coefficient of drag has been examined by Carmichael (1984).

## Slip Flow Regime

Very small particles produce a faster settling velocity than expected from Stokes' law. This is attributed to the slip of fluid molecules at the particle surface, resulting in less resistance encountered. Various correction factors have been applied to Stokes' law to correct for the slip flow effect and to extend the lower particle range. The Cunningham slip factor ($C_c$) extends the Stokes' law region to 0.1 μm (Hinds 1999):

$$C_c = 1 + \frac{2.52\lambda}{d_p} \quad (5.40)$$

where $\lambda$ is the mean free path of gas molecules and $d_p$ is the particle diameter.

The slip correction factor (C) extends the range of Stokes' law to below 0.01 μm (Hinds 1982c):

$$C = 1 + \frac{\lambda}{d_p}\left[2.514 + 0.800 \exp(-0.55\frac{d_p}{\lambda})\right] \quad (5.41)$$

Values of C are close to unity for particles larger than 1 μm. Larger values are observed as the particle size decreases. Tabulated values of C for spheres of unit density are available (Hinds 1982c).

## Centrifugal Settling

The rate of particle settling in a centrifugal field is based on Stokes' law, where the gravitational acceleration is replaced by the centrifugal acceleration (Dallavalle 1948b):

$$v_r = \frac{d^2\omega^2 R(\rho - \rho_0)}{18\eta} \tag{5.42}$$

where $v_r$ is the centrifugal settling velocity; R is the radius of the circular path; $\omega$ is the angular velocity of the particle; $\rho$ is the density of the particle; $\rho_0$ is the density of the fluid; d is the particle diameter; and $\eta$ is the fluid viscosity.

Centrifugation is generally used for sedimentation of extremely fine particles. Particles may be suspended in air or in a liquid medium. Cyclone separators utilize radial acceleration to separate particles suspended in a gas stream (Dallavalle 1948b). The dust-laden gas enters the cyclone tangentially, causing a vortex that descends into the cone. The circular motion imparted on the airstream causes the suspended matter to move toward the outer walls and fall into a collecting bin located below. A greater separating force is required to move smaller particles from the airstream to the outer wall. The tangential velocity of the particle is assumed to be close to that of the airstream.

## Acceleration, Deceleration, and Curvilinear Motion

When a particle undergoes straight-line acceleration under the influence of constant or varying forces, its velocity depends upon the particle mass and mobility. The relaxation time, $\tau$, is the time required for the particle's velocity to adjust to the new condition of forces (Hinds 1982a):

$$\tau = mB \tag{5.43}$$

where m is the particle mass and B is the mechanical mobility of the particle. Constant velocity is considered to be reached when $t = 3\tau$, where t is the time since acceleration/deceleration was applied. Particle motion usually occurs over much longer time periods. However, it may be assumed that a particle reaches terminal velocity instantaneously.

A particle will continue to travel after an external force is removed, depending on the initial velocity and the relaxation time of the particle. The stopping distance (S) in the absence of external forces is given by (Hinds 1982a):

$$S = \tau v_0 = BMv_0 \tag{5.44}$$

where $v_0$ is the initial velocity.

In a moving gas stream passing around an obstacle, the particles follow a curved path. Smaller particles with negligible inertia follow the gas streamlines. Large and heavy particles may impact the obstacle due to inertia. Curvilinear motion is characterized by Stokes' number, which is the ratio of the stopping distance of a particle to the size of the obstacle (Hinds 1982a). As Stokes' number approaches zero, the particles follow the gas streamlines perfectly, and as Stokes' number increases, the particles resist changing their original direction. Collection of particles by inertial impaction is used widely for size fractionation of aerosols, as discussed further in Chapter 6.

## Particle Motion in Liquid Media

Movement of particles within a liquid dispersion occurs by diffusion or gravitational sedimentation, depending on particle size. Field-flow fractionation, a method used to separate particles with liquid media, will be discussed in the following section. Rheology of liquid dispersions will also be discussed in the following section.

### Diffusion

Particles diffuse spontaneously by Brownian motion from regions of higher concentration to regions of lower concentration until equilibrium is reached. Brownian motion is a random movement caused by the bombardment of particles by the liquid molecules. Higher particle velocities and faster diffusion arise with smaller particle size. Brownian motion is reduced by increased viscosity of the dispersion medium. The rate of particle movement due to diffusion is described by Fick's first law of diffusion (Martin 1993c):

$$\frac{dM}{dt} = -DS\frac{dC}{dx} \qquad (5.45)$$

where dM/dt is the rate of mass movement; dC/dx is the concentration gradient; D is the diffusion coefficient; and S is the area of diffusion.

## Gravitational Sedimentation

Stokes' law may be used to describe the rate of gravitational settling for dilute suspensions containing 0.5–2 percent solids concentration. However, Stokes' law does not hold true for suspensions with higher solids concentrations, which exhibit hindered settling because the particles interfere with one another (Martin 1993a).

Brownian motion is more significant with colloidal particles, and gravitational settling is negligible. A stronger force, such as centrifugal acceleration, is required for sedimentation of colloidal dispersions. In coarse suspensions of particles smaller than 2–5 μm, Brownian motion counteracts gravitational sedimentation, producing stable suspensions. Although the particle size is greater than 0.5 μm, these coarse suspensions may exhibit Brownian motion at low viscosity.

## Field-Flow Fractionation

Field-flow fractionation (FFF) is an elution technique capable of simultaneous separation and measurement of macromolecular, colloidal, and particulate materials, ranging from about 1 nm to more than 100 μm. The separation occurs by differential retention in a stream of liquid flowing through a thin channel (50–300 μm) with an applied gradient field at right angles to the flow (Giddings 1993). The applied field drives particles into different stream laminae, and the unequal velocities of the laminae result in the separation of particles along the flow axis. Particles moving in the flow laminae closest to the wall are more slowly entrained. The separated components are eluted one at a time into a detector. The observed retention time is related to various physicochemical properties, such as mass, size, density, and electrical charge.

Sedimentation FFF employs the spinning of the channel to generate differential acceleration forces at right angles to the flow. These forces are highly selective in separating colloidal and larger particles. The single-particle force (F) in sedimentation FFF is:

$$F = V_p \Delta\rho G \tag{5.46}$$

where $V_p$ is the particle volume; $\Delta\rho$ is the difference in density between the particle and carrier liquid; and G is the field strength or acceleration.

Flow FFF employs two right-angle flow streams. The channel stream sweeps components toward the outlet. The crossflow stream, entering and exiting through permeable walls, drives components toward the accumulation wall. The driving force is the viscous force exerted on a particle by a crossflow stream. Other examples of applied fields used in FFF include temperature and electrical fields.

## Flow of Liquid Dispersions

Rheology describes the flow of liquid dispersions. Liquid systems are classified as either Newtonian or non-Newtonian. The rate of shear is proportional to the shearing stress in Newtonian systems (Figure 5.5) according to the following relation (Martin 1993f):

$$\frac{dv}{dr} = \eta \frac{F}{A} \tag{5.47}$$

where $dv/dr$ is the rate of shear; $F/A$ is the shearing stress; and $\eta$ is the coefficient of viscosity, commonly known simply as viscosity, and defined as the resistance of a system to flow under an applied stress. Higher temperature produces increased viscosity of gases and decreased viscosity of liquids. As the viscosity increases, greater applied forces are required to produce flow. The viscosity of colloidal systems depends on particle shape of the dispersed phase. Spherical particles produce lower viscosity than linear particles.

The three types of non-Newtonian flow behavior are plastic, pseudoplastic, and dilatant (Figure 5.5). In plastic flow, flow begins only when a certain shear stress, known as the yield value, is exceeded. At shear stress less than the yield value, elastic deformation occurs. At shear stress greater than the yield value, Newtonian behavior is exhibited. Plastic flow is associated with flocculated particles in concentrated suspensions. Higher yield values indicate a

**Figure 5.5** Rheograms of liquid dispersions. Adapted from Martin (1993f).

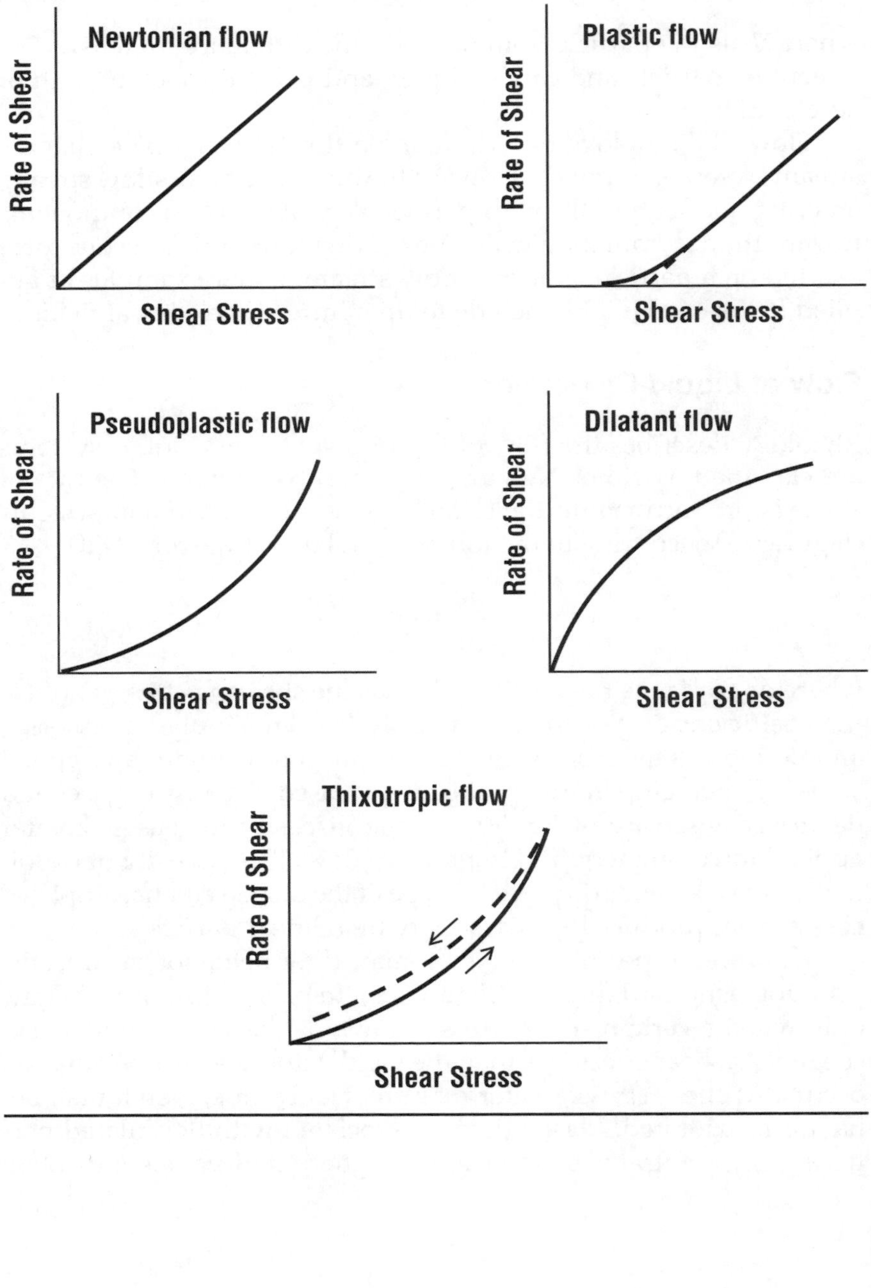

higher degree of flocculation. Systems exhibiting plastic flow may be characterized by using the yield value and the plastic viscosity. Materials exhibiting plastic flow are known as Bingham bodies.

In pseudoplastic flow, increasing shear stress produces an increased rate of shear and a reduced viscosity. Pseudoplastic flow is exhibited by polymers, due to linear alignment of molecules with increasing shear stress. This alignment reduces the internal resistance and increases the shear rate. Pseudoplastic systems are also known as "shear-thinning" systems.

In dilatant flow, increasing shear stress produces an increase in viscosity. Dilatant systems increase in volume under applied shear force. Dilatant systems are also known as "shear-thickening" systems. When the shear stress is removed, the system returns to its original state of fluidity. Dilatant systems generally contain a high solids concentration (>50 percent) of small-deflocculated particles. At rest, particles are closely packed with minimum void space filled with fluid. At increasing shear stress, the particles need to take an open packing form to move past each other, resulting in expansion or dilation of the system. The volume of fluid then becomes insufficient to fill the increased void space. Increased resistance to flow results because the particles are no longer completely wetted or lubricated by the fluid.

The removal of shear stress produces a hysteresis in the rheograms of plastic and pseudoplastic systems, due to the breakdown of structure under shear stress. The restoration of structure occurs with time. This phenomenon is known as thixotropy. The rheogram of a thixotropic system depends upon the rate of applied and removed shear stress and the length of time the sample is subjected to the stress. Thixotropic suspensions are desirable for pharmaceutical formulations. These suspensions do not settle readily during storage, become fluid on shaking, remain in a suspended state as the dose is dispensed, and regain consistency for further storage (Martin 1993f).

## Conclusion

The characterization of fundamental particle properties is discussed in this chapter. Particle size is generally described in terms of equivalent spherical diameters. Particle shape may be characterized by using geometric shape factors or mathematical approaches, such as Fourier and fractal analyses. Surface roughness is generally determined by profilometry and may be characterized by using parameters such as rugosity, root mean square deviation, and maximum peak-to-valley height. The surface area of particles is most commonly determined by the BET nitrogen adsorption method. The bulk density of powders depends on the true density and packing arrangement of particles, which in turn is affected by the particle size distribution, shape, and adhesional properties.

Particle behavior is examined with particular focus on particle adhesion and motion within bulk powders, gaseous dispersions, and liquid dispersions. Electrostatic charge measurements are performed using the Faraday well or cage for dry powders, or electrophoresis for liquid dispersion systems. Adhesion force techniques measure the detachment of either single or multiple particles. Single-particle detachment is measured by using microbalance, pendulum, or atomic force microscopy methods. Multiple-particle detachment is measured by using centrifugation, vibration, impaction, and aerodynamic techniques. Fundamental particle properties, such as chemical composition, particle size, particle shape, and surface roughness, greatly affect particle adhesion. Environmental conditions, including storage duration and humidity, also affect particle adhesion. Blending of powders may involve the random mixing of noninteracting particles or may utilize the adhesion of particles, known as interactive mixing.

Powder flow may be characterized by using static bed measurements, such as the angle of repose or spatula, Carr's compressibility index, the Hausner ratio, or shear cell measurements, or dynamic flow measurements, including the rotating drum or vibrating spatula. Interparticulate forces need to be overcome for powder flow to commence, so factors affecting powder flow are similar to those affecting particle adhesion. The gas fluidization of powders is classified according to Geldart (1986). Particle motion in gaseous and liquid dispersions involves either diffusion by Brownian motion or gravitational sedimentation. Stokes' law describes the rate of particle settling in laminar conditions under gravity. Slip correction factors are applied to Stokes' law for very small particles. A variation of Stokes' law enables calculation of the rate of particle settling in a centrifugal field.

The rheological behavior of liquid dispersions is determined from the rate of shear as a function of shearing stress. The flow behavior is described as Newtonian and non-Newtonian, which includes plastic, pseudoplastic, and dilatant behavior. Plastic and pseuodoplastic systems may exhibit thixotropy.

Fundamental particle properties have a great influence on particle behavior. An adequate and specific characterization of pharmaceutical particulate systems is required to understand particle behavior.

# 6

# Instrumental Analysis

The principles behind various methods employed to assess particle morphology are embodied in myriad commercially available instruments. Their operation is illustrated in this chapter, although details of particular instruments will not be described. Particle size and morphology measurement may be divided into four classes: direct imaging, indirect imaging, physical methods, and charge analysis. A number of different instruments lie within these broad categories, each with its own particular strengths and limitations. Consideration of approaches to calibration of particle sizing instrumentation is also presented. Finally, comments are made on selection criteria for a particular method or group of methods. The average dimension being measured is the most important criterion for choosing any instrument. Figure 6.1 presents a sketch of the operational range of particle sizes for the methods described in this chapter. The significant overlap among the techniques provides the opportunity to utilize more than one method to develop a more complete picture of the overall product.

**Figure 6.1** Range of operation for particle sizing methods.

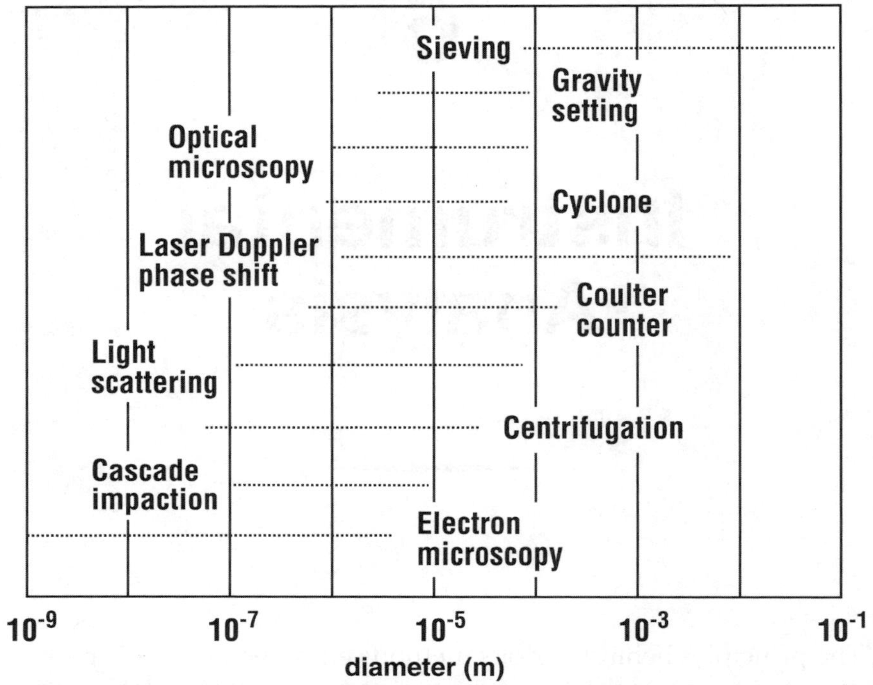


## Direct Imaging

Direct imaging is an absolute method of particle size and morphology characterization because it is the only technique in which the instrument itself does not provide some interpretation of the data. Direct imaging is also the only method that provides the ability to examine individual particles rather than a population of particles. Microscopy, a direct-imaging technique, provides a confirmation for any other method of particle sizing. It also provides the ability to identify irregular particle morphology as well as aggregates, hollow particles, or other forms of nonhomogeneity. Sizing particles in a microscope image involves comparing the particles with a reference, either in real time or from a captured image. Therefore, some interpretation is required in direct imaging because the reference comparison is operator dependent. In addition, for nonspherical particles, using various equivalent diameter expressions, as described in Chapter 4, introduces further errors of interpretation. Direct imaging includes the methods of optical, electron, and atomic force microscopy, described next.

## Optical Microscopy

Optical microscopy uses visible light to examine particles. The wavelength of light determines the fundamental limit of resolution for this method. The resolving power of a microscope is equal to the wavelength of illuminating light divided by the sum of the numerical aperture of the objective lens and the numerical aperture of the condenser. The numerical aperture is determined by the index of refraction between the object and the lens and the half-angle of light entering the lens. There is a trade-off, however, between resolution and depth of focus: increasing the numerical aperture increases resolution but decreases the depth of focus. In general, the highest focal length (depth of focus) suitable for a given sample should be employed. Optical microscopy can, in theory, be used to image particles larger than about 1 μm. Practically, however, difficulties in depth of focus at high magnifications limit the utility of light microscopy for particles sized near the resolution limit. Electron microscopy, discussed in a later section, is the method of choice for sizing of particles near, and somewhat above, the light microscopy resolution limit.

Microscopy requires the preparation of a slide containing a representative sample of uniformly dispersed powder. Most methods of preparing the slide involve dispersing the sample as a suspension in a liquid and then spreading a drop of the fluid on the slide (Allen 1968). The result of this procedure is, unfortunately, largely operator dependent. The powder-containing fluid can be spread either by centrifugal force or by physical means such as a glass rod. Care must be taken when spreading powders to not fracture the particles. The goal in spreading the particles is to achieve a low density of particles per unit area (Kaye 1981). High densities can result in judgment errors in sizing because the size of nearby particles in the image may influence the size determination made by the operator.

Determination of particle sizes from microscope images may occur in real time or from a captured image. In both cases, a reference scale placed in the image is compared to particles in the image. An older method of measuring particle sizes uses a microscope equipped with an optical micrometer linked to a movable crosshair in the ocular. The crosshair is moved to the edge of a particle, the micrometer reading is recorded, and the crosshair is then moved to the opposite edge. The difference in micrometer readings between opposite edges is the particle diameter. This is clearly a very time consuming technique and has been supplanted by other methods.

Particles may be sized by ocular scales or graticules. In each case, the slide is placed on a stage that can be moved in perpendicular directions. The slide is examined in strips, and the size of the particles passing over the scale in the image is recorded. Measurement is performed in vertical or horizontal strips to provide thorough coverage of the slide. The diameter of nonspherical particles is taken in the direction of the scan with the idea that differences in orientation of particles will average out over the whole image. The number of particles that must be counted to achieve a representative sample of the overall particle size distribution is discussed in Chapter 3. Particle size determination by graticules involves the comparison of individual particles with circles of decreasing diameter. Most graticules use a $\sqrt{2}^n$ progression in circle diameter, where n is the nth circle. This progression is used because many aerosol systems studied with these graticules are distributed lognormally, and the $\sqrt{2}$ progression gives equally spaced data points when plotted on logarithmic graph paper. The Patterson Globe and Circle or Porton graticules are examples of this type of graticule. Graticules should be calibrated before use with a stage micrometer.

Achieving a representative sample of the overall particle size distribution clearly requires sampling a large number of particles in the image. Various sources recommend differing counts. Allen (1968) suggests that as few as 600 particles may be counted. The ASTM method requires measurement of 10 particles in the largest and smallest size classes in the tails of the distribution (ASTM 1973). This may not be practical for a very narrow particle size distribution because, statistically, so few particles would be found in the tails that very large sample sizes would have to be used.

Microscope images may be acquired digitally and sized with an automated or semi-automated approach. It should be noted that the combination of photography and optical imaging could lead to errors in measurement because weight is given to particles in focus. In practice, a number of fields of view should be used for statistical validity. In the semiautomated method, an operator uses the computer mouse to draw reference lines across the particles. The software calculates the diameters from the length of a calibration line in the image whose length is entered by the user. Digital acquisition offers the further advantage of readily available archiving and reproduction of the image data. Automated systems are also available. There is a potential for error in these systems, however, especially if the distribution of powder in the image is too dense. The software may view two overlapping particles as a single particle. The semiautomated method combines ease of use with the ability to include human judgment in the process.

Optical microscopy is the least expensive and most readily available method of particle sizing. Drawbacks to the microscope include the potential for operator error, especially given the tedium of particle sizing by this method. Additionally, for significant statistical representation of a size distribution, at least 300 particles should be counted. Finally, this method does not offer the ability to separate particles by size.

## Polarized Optical Microscopy

Polarized light microscopes add two polarizing filters to an ordinary optical microscope—one below the microscope stage to polarize illuminating light and one situated between the objective lens and the eyepiece. In addition to the polarizing filters, the microscope should be fitted with a rotating sample stage and a crosswire graticule to mark the center of the field of view. Polarized light microscopy provides the ability to distinguish between isotropic and anisotropic materials. Isotropic materials demonstrate the same optical properties in all directions, whereas anisotropic materials have optical properties that vary with the orientation of incident light with the crystal axes.

In the typical setup, the two polarizing filters are oriented with their polarization directions at right angles so that, in the absence of a sample, no light passes through the system and the field of view is dark. A sample placed on the stage is rotated through 360 degrees. With the crossed polarizers, anisotropic materials can be distinguished from isotropic materials because they will be illuminated at some angle during rotation, whereas the isotropic materials will remain dark. In another use, the analyzer polarizer can be rotated out of the optical path. Anisotropic materials will change color when rotated because their absorption color varies with the orientation of the crystal structure to the polarized incident light. Polarizing microscopy has been used to examine the crystal habit and particle size of crystalline drugs by Watanabe (1997).

## Dark-Field Microscopy

As with light microscopy, dark-field microscopy creates a contrast between the object being imaged and the surrounding field. However, as the name implies, in dark-field microscopy the image background is dark and the object is bright. This technique is particularly useful for visualizing samples that have refractive indices very close to those of their surroundings (Firestone et al. 1998).

In conventional light microscopy, an image of the sample is formed due to reflection of light off the sample or due to refraction of light from changes in the refractive index as light strikes the sample. Both reflection and refraction produce small angular changes in the direction of the light ray. In the case of a sample with a low index of refraction, the angular changes are so small as to be overwhelmed by the unrefracted light from the illumination source. The dark field is produced by an annular stop placed in the path of the illuminating light before it reaches the specimen (Gray 1964). Almost any laboratory microscope can be converted to a dark-field microscope by adding an opaque stop. With the stop in place, only oblique rays pass through the specimen on the microscope slide (see Figure 6.2). With no specimen, an oblique hollow light cone is formed in the direction of the objective lens. If the numerical aperture of the condenser is greater than the objective, the oblique rays will fail to enter the objective and the field will appear dark. Reflection, refraction, and diffraction from discontinuities in the sample will alter the path of incident light so that it is collected by the objective. The result is an illuminated image in a dark field. In Fourier

**Figure 6.2** Dark-field microscopy setup.

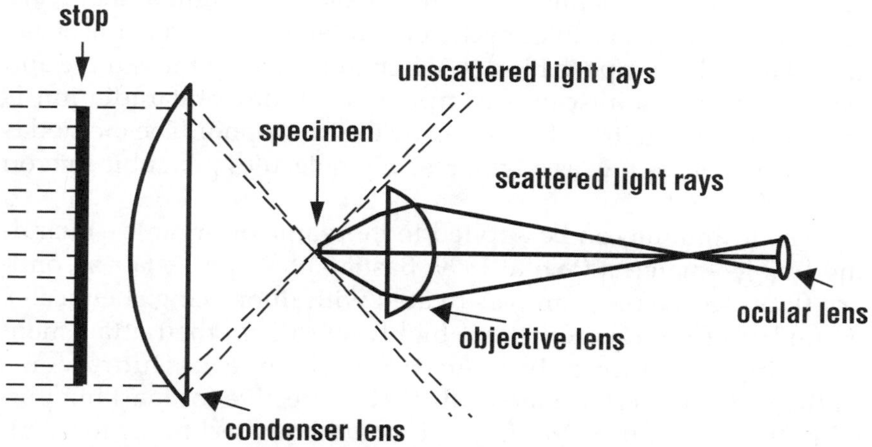


terms, the annular stop serves to remove the zero order (i.e., the unscattered light) from the diffraction pattern, leaving only higher-order diffraction intensities to be collected by the objective.

Impurities such as dust will scatter light as effectively as the sample, so much care must be taken when preparing slides for dark-field microscopy. Specimen thickness is an important consideration. A thin sample is preferable, to minimize diffraction artifacts that can interfere with the image. Microscope slides will have the same effect and therefore should be no more than 1 mm thick. Because the annular stop eliminates bright, undeviated light, a high-intensity light source must be used (Davidson 1995).

## Fluorescence Microscopy

Fluorescence refers to the reemission of absorbed light at longer wavelengths than the excitation light. This occurs when a photon of excitation radiation collides with an electron in the atomic cloud, exciting it to a higher energy level. The excited electron then relaxes to a lower level and, because energy must be conserved, emits the light as a lower-energy photon. Fluorescence microscopy takes advantage of this effect to image materials that fluoresce either naturally or when treated with fluorescing chemicals. Fluorescence microscopes excite the sample with ultraviolet (UV) light and observe emitted light in the visible spectrum. A filter is required in the objective of the microscope to block UV light reflected by the sample and pass only the fluorescence component. Fluorescent radiation is spherically symmetrically emitted by the fluorophor. The method is very sensitive; as few as 50 fluorescing molecules per cubic micron can be detected.

The technique can be applied to inorganic or organic—including living—material (White 1999; Bastian et al. 1998). For secondary fluorescence (i.e., samples treated with fluorescing chemicals), a number of fluorochromes are highly specific in their attachment targeting. A commonly used fluorochrome is fluorescein-isothiocyanate (FITC), a molecule that is effective for marking proteins. It is widely used in a fluorescent antibody technique for identification of pathogens. An interesting application of this microscopy is the visualization of particle uptake by macrophages (Koval et al. 1998). Fluorescence microscopy provides the ability to view this process in real time.

## Transmission Electron Microscopy

The fundamental limit of resolution imposed by the wavelength of light prompted the development of the transmission electron microscope (TEM). Decreasing the wavelength through the electromagnetic spectrum would suggest the use of UV or X-ray wavelengths for microscopy. Unfortunately, UV light is absorbed by glass, so UV microscope optics must be constructed of quartz. X-ray microscopes have been constructed but require extremely intense sources, such as synchrotrons, and complicated optics, such as Fresnel zone plates (Cosslett and Nixon 1960).

The wave/particle duality of electrons provided an alternative source. Electrons can be accelerated by an electrical potential and refracted in a magnetic field. The faster the velocity of the electron, the shorter the wavelength of the electron wave, as given by the de Broglie equation:

$$\lambda = \frac{h}{mv} \tag{6.1}$$

where h is Planck's constant; m is the electron mass; and v is the velocity. Transmission electron microscopes offer approximately a thousandfold increase in resolution and a hundredfold increase in depth of field.

A drawback of the TEM is that electrons are quickly absorbed by any matter they encounter. Therefore, TEM imaging must occur under high vacuum to minimize electron collisions with air molecules. The sample being imaged must be sectioned extremely thin for the electrons to penetrate the material. Finally, the TEM stage must be scanned through the electron beam. The scanning electron microscope (SEM) addresses these limitations and is of greater utility to the pharmaceutical scientist.

## Scanning Electron Microscopy

The scanning electron microscope (SEM) provides a very versatile and powerful imaging ability. Resolutions on the order of 2 to 5 nm or even smaller have been obtained by commercial instruments (Goldstein et al. 1992). In addition to their high resolution, an important feature of SEM images is the three-dimensional appearance of the images. This comes about due to the large depth of field achieved by the instrument. This depth of field may be one of the most valuable features of this method of microscopy.

The typical commercial SEM consists of a lens system, electron gun, electron collector, cathode ray tubes (CRTs) for displaying the image, and electronics for digital storage of the image. The SEM operates by scanning a fine beam of electrons of medium energy (5–50 keV) in parallel tracks across the sample. The collision of electrons with the sample material results in the generation of secondary electrons, Auger electrons, backscattered electrons, light or cathodoluminescence, and X-ray emission. Samples of up to 25 × 25 mm can be imaged at magnifications from 20× up to 100,000×.

The ultimate resolution of the SEM depends on the electron beam diameter, and the intensity of emitted signals depends on the beam current. These two parameters operate in opposition to each other because a small beam can carry a smaller current than a larger beam. Therefore, the operator must balance these controls to produce the desired image result. The electron beam produced by the electron gun is too large to produce a sharp image. Electron lenses are used to focus the beam into a desirable spot size. The SEM chamber is operated under vacuum to minimize scattering of the electron beam by gas molecules in the chamber. Evacuation of the sample chamber necessitates special preparation of the sample.

The electron beam is scanned across the sample with two pair of scan coils that deflect the beam. The sample signals most often used are secondary electrons and backscattered electrons. These electrons are collected by an Everhart-Thornley (E-T) electron detector. The E-T detector is electrically isolated from the chamber and has a positively charged wire mesh screen at its entrance. The positive charge attracts the low-energy secondary electrons to the detector. Once they enter the detector, the electrons are accelerated and strike a scintillator material, which produces light that is amplified by a photomultiplier tube.

Preparation of organic materials for SEM imaging requires a trade-off between instrument performance and sample integrity. Organic materials are typically hydrated and composed of elements with low atomic numbers. They may have low thermal and electrical conductivity. Thus, these samples are prone to damage when confronted with a high-energy electron beam. The samples must be dried and coated with conductive material if they are not to compromise the operation of the SEM. An environmental SEM at higher pressures, as discussed in the next section, can be used when sample treatment is not possible. Samples can be dried by chemical dehydration, freeze-drying, critical point drying, or ambient-temperature sublimation. The samples must then be coated with a conducting material to reduce electric charge buildup on the samples. Charging of the sample may cause dielectric breakdown of the sample, thus altering its morphology. Charged areas of a sample may degrade the image resolution by deflection of low-energy secondary electrons or deflection of the electron beam. Samples are typically made conductive by sputter coating with a thin layer of gold or palladium. Heating of the sample from the low-energy electron beam is not typically an issue but could alter heat-labile samples.

## Environmental Scanning Electron Microscopy

The SEM offers many advantages, as discussed in the previous section. A drawback to the conventional SEM, however, is that a significant amount of sample preparation is required prior to imaging. In many cases, sample preparation may simply be an inconvenience, but in some cases, the freezing, drying, and coating processes may alter the surface morphology of the sample. The environmental SEM (ESEM) provides the ability to image samples in their wet state, eliminating many of the SEM preparation steps.

Environmental SEMs operate at pressures ranging from 0.1 to 1 kPa. The electron column remains under vacuum through the use of pressure-limiting apertures within the lens (Danilatos 1991). Some samples can be imaged directly by using the ESEM, whereas others do require some care in the performance in the instrument. Sample hydration in the ESEM chamber must be monitored as the chamber is prepared. The ESEM chamber can be flooded with saturated vapor during pumpdown to insure continued hydration of the sample. A number of pharmaceutical specimens have been successfully imaged by the ESEM, including biodegradable polymeric matrices (polyanhydrides and lactide-co-glycolide copolymers), hydrogels, and pellet-based formulations (D'Emanuele and Gilpin 1996). In the case of pellet formulations, the ESEM has allowed imaging of hydrated film coatings of pharmaceutical pellets. In contrast, the conventional SEM would not allow imaging of coating films in their hydrated state.

## Scanning Tunneling Microscopy and Atomic Force Microscopy

The scanning tunneling microscope (STM) and the atomic force microscope (AFM) operate on the same quantum mechanical principle. The 1986 Nobel Prize in Physics was awarded to Gert Binnig and Heinrich Rohrer for the development of the STM as well as to Ernst Ruska for work on the electron microscope. In the STM, a pointed metal probe is placed within a few nanometers of the sample. With a small (10 mV) voltage applied to the probe, electron tunneling can occur between the probe and the sample. An exponential relationship exists between the separation between the probe and sample and the tunneling probability such that changes in separation as small as 0.01 nm can result in measurable changes in tunneling current. By scanning the probe over the sample surface, bumps on the surface can be measured and a topographic map generated. Piezoelectric ceramic transducers are used to perform the scan because of their precise ability to position the probe. The STM requires that the sample be conductive (Hornyak et al. 1998).

A variation of the STM, the atomic force microscope can be used with insulators. The AFM has been used to characterize surface roughness of pharmaceutical particles (Li and Park 1998) and elastic properties of aerogels (Stark et al. 1998). In an interesting application of the AFM, probe tips were mounted with spherical colloidal silica particles, and the forces of interaction between the colloid probe and other pharmaceutical materials were measured (Fuji et al. 1999).

### Conclusions

Direct imaging of particles provides an irrefutable determination of their size and morphology. Although errors can be made in sizing or morphologies can be altered by sample preparation, direct imaging requires less interpretation of the data than any other method. Some form of direct imaging should be used in combination with any other particle sizing methods to confirm the outcome of those measurements.

## Indirect Imaging

Indirect imaging can be used to measure the actions of a single particle at a time or to collect data from all particles in a sample at the same time.

### Single Particles

Single-particle methods of indirect imaging consist of means of measuring the actions of a single particle at a time. Individual measurements of multiple particles clearly must be taken to develop a distribution. Single-particle techniques do not measure the entire powder or aerosol. Thus, considerations must be made concerning the validity of sampling. Techniques that use single-particle methods include electrical resistance, light blockage, and laser velocimetry.

### *Electrical Resistance*

Electrical resistance (electrozone) particle-sizing methods are used with particles suspended in an electrolyte solution. The Coulter counter, a well-known version of this type of instrument, was developed originally for blood cell counts. In its operation, particles in the electrolyte solution are forced to flow through a small orifice. Electrodes placed on either side of the orifice maintain constant current flow through the orifice. The passage of a particle through the orifice causes the resistivity to increase as the particle displaces electrolyte. This resistivity increase results in a voltage spike in which the amplitude of the spike is proportional to the volume of the particle. Pulses are classified by area, and a count is incremented in a bin corresponding to the size. Counting of multiple individual particles results in a distribution.

The electrozone technique has a dynamic range of sizes from 0.5 to 800 μm, but different orifice sizes must be used to achieve this range. A given orifice can be used with particles ranging in size from around 2 to 50 percent of the orifice size; blockage of the orifice becomes a problem with larger particles, and smaller particles are undetectable above electronic noise. The resolution decreases near the lower limit of the measurable range. The principle of operation is based on displacement of fluid, so the particle volume will be measured regardless of shape. The instrument is easily calibrated with polystyrene size standards. If two particles pass through the orifice at the same time, however, they will be counted as a single particle (Wynn and Hounslow 1997). Other drawbacks include the requirements that samples must be suspended in conductive fluids and must be insulating materials.

### *Light Blockage*

The light-blockage technique of particle sizing is the optical equivalent of the electrical resistance method. In a light-blockage instrument, particles in liquid are passed through a counting chamber, where they partially block a laser beam focused on the chamber. The reduction in light intensity measured by a detector is proportional to the optical cross-section of the particle. Like the electrozone counter, a population of single particles is measured.

This technique is applicable to a broad range of particle sizes spanning from 0.5 to about 2,000 μm, although different size sensors must be used. Analysis times are rapid, on the order of 3 minutes or less. Resolution is decreased for nonspherical particles because the particle cross section is evaluated. For nonspherical particles, the orientation of the particle as it passes through the detector will also have an effect on the measured size.

### *Laser Velocimetry or Time-of-Flight Counter*

The time-of-flight counter, or laser velocimetry method, determines particle size by measuring transit times of individual particles between the beams of two lasers. The technique is used most often with dry powders, although it may be used with nebulized suspensions. Laser velocimeters operate by accelerating particles through a jet or nozzle into a partial vacuum. The particles accelerate according to their size, with smaller particles accelerating faster than larger particles. When a particle passes through the first laser beam, a time-of-flight clock is started and then stopped upon the particle's passage through a second beam. Particle size is calculated from the measured time-of-flight.

The time-of-flight technique is capable of measuring particle sizes ranging from about 0.2 to 700 μm, although resolution is limited at the bottom of the range. The time-of-flight counter can be calibrated readily with solid particle size standards. Measurement times are rapid, normally about a minute. Some instruments are capable of measuring on the order of 100,000 particles per second (Niven 1993). Care must be taken to introduce the appropriate concentration of particles because the counting speed of the instrument is limited (Thornburg, Cooper, and Leith 1999). Nonspherical particles will not be sized correctly, but the magnitude of the error is difficult to determine because it depends on shape factors (Kaye et al. 1999). Due to the high shear forces that are generated, particles are deaggregated by the instrument (Hindle and Byron 1995). Therefore, sizing results may not be relevant to powders that tend to aggregate. In addition, the particle density must be known for use in the particle size calculation.

## Populations of Particles

Ensemble methods of particle sizing collect data from all particles in a sample at the same time and then calculate size distributions from the data. Most of these instruments rely on the Mie theory of light scattering in their sizing algorithms. Ensemble particle-sizing techniques include light scattering, laser diffraction, and ultrasound scattering.

### *Light Scattering*

Light scattering, or photon correlation spectroscopy (PCS), can be used for ensemble sizing of small particles. PCS operates on the principle that light scattered from small particles varies in intensity due to the Brownian motion of the particles. The technique is applicable only to particles for which movement due to Brownian motion is more significant than movement due to sedimentation, such as particles smaller than 3 μm. In this method, particles are dispersed in a liquid and placed in a sample holder that is illuminated by a laser beam (Van Drunen, Scarlett, and Marijnissen 1994). A photomultiplier tube mounted off axis from the laser measures scattering intensity. Brownian motion of the particles causes fluctuations in scattering intensity. This intensity is autocorrelated over short time scales to obtain the diffusion coefficient for the distribution, which is inversely proportional to the decay time of light scattering. The diffusion coefficient, D, is related to the particle diameter, d, by the Stokes-Einstein equation:

$$d = \frac{k_B T}{3\pi\eta D} \tag{6.2}$$

where $k_B$ is the Boltzmann constant, T is temperature, and $\eta$ is the fluid viscosity.

PCS can be used with very small samples of material. Due to the short time scales of Brownian motion, analysis is very rapid. The temperature of the sample must be precisely monitored because the diameter calculation depends on T.

### *Laser Diffraction*

Most laser diffraction particle sizing is based on Fraunhofer diffraction, a special case of the Mie theory (Ma et al. 2000), although some instruments use Mie theory. Fraunhofer diffraction calculations provide the volume of particles given their diffraction pattern. In this sizing technique, laser diffraction instruments pass a laser beam through a sample dispersed in liquid or in air (Ranucci 1992). The focused laser beam is diffracted by particles in the beam, more broadly by smaller particles and less so by larger particles. The scattered light is focused by a lens onto a ring array diode detector that passes scattering intensity data to a computer for particle size distribution determination.

Early laser diffraction instruments used a series of lenses to measure ranges of particle sizes. Instruments currently available can measure a very broad range of particle sizes. In either case, laser diffraction can measure a dynamic range of particle sizes from 0.5 to 1800 μm. Most laser diffraction instruments have accessories for measuring dry powders dispersed as aerosols and nebulized solutions or suspensions. They can be arranged for synchronization of the metered dose inhaler spray and the data collection. Alternatively, the particles may be suspended in a liquid for measurement.

Laser diffraction instruments have been widely used through the past few years, primarily because of the ease of measurement. Data collection with this technique is very rapid. Data are expressed in terms of volume distributions, although other parameters may be calculated. Errors due to multiple scattering may occur if the concentration of particles is too large. Multiple scattering is the effect by which light scattered from one particle is scattered by a second particle before reaching the detectors. The particle concentration in the laser beam must be chosen to provide sufficient diffracted light but not be high enough to cause significant multiple scattering events. The index of refraction of the particle must differ from the suspending medium (air or liquid). Latex spheres of stated size guaranteed by the manufacturer may be purchased for calibration. It is also possible to purchase a reference reticle from the instrument manufacturer. This reticle has a polydisperse particle size distribution of chrome dots on quartz with a diffraction pattern determined by the manufacturer from a properly calibrated and aligned instrument (Ranucci 1992).

Particle size distributions from laser diffraction instruments are based on calculations from theory. All instruments base this calculation on Fraunhofer or Mie diffraction theory, but the precise algorithm is typically considered proprietary and not provided to the user. This leads to the tendency to treat these instruments as "black boxes." Another technique, such as microscopy, should always be used to confirm light scattering results.

### *Ultrasound Sizing*

Ultrasonic energy can be used to perform size and chemical characterization of particles in liquid (Povey 2000). Acoustic spectrometry calculates particle sizes from scattering of ultrasound energy in exactly the same manner as particle sizing is determined by using light scattering. A particular advantage of ultrasound characterization is the ability to measure particle size in liquids, including opaque dispersions. Acoustic spectroscopy has a dynamic range of 10 nm to 1 mm. As such, the technique can be used for in-line process monitoring. A drawback to acoustic spectroscopy is the long acquisition times (up to 3 minutes) required to take a full sample measurement.

### *Gas Adsorption*

Gas adsorption studies can be used to determine the surface area of an adsorbing solid, as discussed in Chapter 5. Recent interest in porous particles for aerosol delivery makes gas adsorption an especially important technique (Edwards et al. 1997). If it is assumed that particles are spherical, the volume-surface diameter can be calculated from the surface area (Martin 1993). If $\rho$ is the density of the adsorbent material, and $S_w$ is its calculated specific surface area in square centimeters per gram, then the volume diameter, $d_v$, is given by:

$$d_v = \frac{6}{\rho S_w} \tag{6.3}$$

## Physical Methods of Particle Size Separation

Physical methods of particle size separation take advantage of fundamental physical properties of particle dynamics to segregate particles by size. Most of the methods described in this section utilize the settling properties of particles in a gravitational or other force gradient field. Some of the techniques discussed offer the potential to collect size segregated particles for further use. Others, such as the impaction methods, provide information relevant to particular states of the powder, that is, the aerodynamic properties of the powder dispersed as an aerosol.

### Sieving

Sieving is a method of particle size analysis that can also provide size-segregated particles for use in processing. The typical sieve is a cylindrical open container constructed with a base of wire mesh of known opening size. For particle sizing, sieves are stacked in increasing aperture size. Powders are placed in the top sieve, then covered and agitated. A closed collection pan is placed below the sieve of smallest size to collect the fines. After agitating for a sufficient time for the smallest particles to settle through the stack, the contents of each sieve stage are recovered and weighed. Results are presented at cumulative percent undersize for the aperture diameter of the given stage (Kaye 1981).

Sieves are supplied according to a standard, such as the British Sieve or the International Organization for Standardization (ISO) standard. The U.S. standard is a $\sqrt{2}$ progression of sizes starting at 350 μm down to 45 μm. Sieves are also classified by their mesh size, measured by the number of wires per inch. Sieving results may vary as a function of sieve load, particle shape, agitation method, physical properties, and overall size distribution. For example, Hickey and Concessio (1994) have noted that some pharmaceutical powders tend to aggregate when sieved by vibration versus tapping. Other errors may occur due to errors in sieve construction or due to wear. In general, if losses in sieving exceed 0.5 percent of the original sample weight, the data should be discarded.

## Inertial Impaction

Inertial impaction is most frequently used in pharmaceutical science to size particles intended for aerosol dispersion. It is also used extensively in environmental science to sample naturally occurring or human-produced aerosols such as pollutants. Inertial impaction utilizes particle inertia to segregate aerosol particles into discrete size ranges (Hinds 1982b). Impactors were initially developed in the 1940s to analyze environmental aerosols (May 1982). These devices consist of circular orifices of decreasing diameter arranged on stages in series. An impaction plate is positioned beneath each orifice stage. The airflow velocity through the orifices increases as their diameter decreases because the system is closed and the air volume flow rate drawn through the impactor is constant. If aerosol particles are to remain entrained in the airflow through the device, they must follow the curving air streamline through the orifice and around the impaction plate. Particles with an inertia too great to achieve the angular acceleration required to remain in the streamline will continue in their initial direction and impact on the collection plate. Particles with a smaller aerodynamic mass, and thus a smaller inertia, will continue to the next stage of the impactor. The process continues through the impactor, and the particles are segregated by size and inertia as the airflow velocity increases through the narrowing jets.

The particle size measured by inertial impaction is the aerodynamic size. Gravitational and aerodynamic forces on particles were discussed in Chapter 5. The aerodynamic size of a particle is a function of the balance of aerodynamic and gravitational forces on a particle, given its size and density. The aerodynamic diameter of a given particle is defined as that of a unit density sphere with an equivalent settling velocity, such that:

$$\rho_a d_a^2 = \rho d^2 \tag{6.4}$$

where $\rho_a$ and $\rho$ are the aerodynamic and true densities of the particle, respectively, and $d_a$ and d are the aerodynamic and geometric diameters of the particle, respectively. Since $\rho_a = 1\ g/cm^3$, the equation relating the geometric diameter to the aerodynamic diameter for spherical particles is:

$$d_a = \sqrt{\rho}d \tag{6.5}$$

For nonspherical particles, a shape factor must be included in equations (6.4) and (6.5). A shape factor ranging from approximately 1.1 to 1.75 is frequently used as a multiplicative term on the right-hand side of equation 6.5. The practical and theoretical implications of shape on aerodynamic properties have been discussed by many authors (see, for example, Reist 1993a; Colbeck 1990; Fults, Miller, and Hickey 1997).

Inertial impaction is the particle sizing method prescribed by compendial specifications for medical aerosols (BP 1999; USP 2000). Its utility stems from three major factors. First, the entire aerosol generated by an inhalation device can be sampled by inertial impaction. Second, the collection of particles in the impactor is assayed by chemical means so the size distribution can be calculated for drug particles only. Finally, the size measured by impactors is the aerodynamic size. The aerodynamic size is the most relevant to inhalation because this is the size that influences deposition in the lungs.

When sizing aerosols to be generated by inhalation devices, care must be taken that the aerosol is representative of the aerosol that would be generated by the device in clinical use. This specification may conflict with the very specific operational parameters under which impactors are calibrated. For example, the Andersen impactor is calibrated at flow rates of 28.3, 60, and 90 liters per minute (Lpm); a 28.3 Lpm flow rate may not be sufficient for dispersion from a dry powder inhaler (DPI), but 60 Lpm may not be achievable by certain patients (Clark and Hollingworth 1993). Another drawback to inertial sizing is that collection of the particles must be performed manually. The procedure is very labor intensive and time consuming. Finally, impactors distribute an aerosol into discrete size ranges, typically ranging from 0.2 to 10 μm. A larger range of sizes or finer resolution might be preferable for some applications, such as sizing of nasal sprays.

Of the various impactor types available, the impactor given as an example in the USP is the Andersen Mark II nonviable 1 ACFM ambient sampler (USP 2000). This impactor has eight stages and is calibrated for routine use at 28.3 Lpm (1 ACFM). At this flow rate, particles are separated into fractions of >9.0, 5.8, 4.7, 3.3, 2.1, 1.1, 0.7, and 0.3 μm. Particles that do not impact on the last stage are collected in a backup filter with a pore size of 0.2 µm. The Andersen impactor is also available with modified initial stages calibrated for use at 60 and 90 Lpm. At 60 Lpm, the cutoff diameters for the stages are 8.6, 6.2, 4.0, 3.2, 2.3, 1.4, 0.8, and 0.5 µm. When operated to collect dry powder aerosols, the impactor is equipped with a preseparator stage that has a cutoff diameter of 9.9 µm. The preseparator is designed to remove very large particles from the aerosol cloud. These particles would be subject to bounce and reentrainment and could overwhelm the first stage of the impactor. Many authors have evaluated the Andersen impactor for the accuracy of its calibration, for errors in manufacture, and for errors in operation (see, for example, Vaughn 1989; Swanson et al. 1996; Stein 1999).

Errors in particle sizing using the Andersen impactor, and most other impactors, result from particle bounce and wall losses (Vaughn 1989; Mitchell, Costa, and Waters 1988). Particle bounce occurs when particles, especially dry particles, are not captured by the collection plate with which they collide, and instead are reentrained in the airstream and collected at a lower stage. Particle bounce can be minimized by coating the impaction plates with silicon oil (Esmen and Lee 1980). Wall losses (interstage losses) in impactors are a size-dependent phenomenon wherein particles impact on the walls of the impactor as they transit from one stage to the next or before moving through a jet (Mitchell, Costa, and Waters 1988). Wall losses cannot be eliminated. The USP specifies that if interstage losses account for more than 5 percent of the total dose from a device, the interstage mass can be included with the associated impaction plate in size distribution calculations.

Another type of inertial impaction instrument is the liquid impinger. Impingers use a liquid as the inertial impaction surface. As such, they are well suited for use with powder aerosols because impinging particles are captured by the liquid, thereby eliminating the possibility of particle bounce. In addition, sample collection from liquid impingers is less tedious than for impactors; the liquid is removed with a pipette and analyzed. Commercially available impingers have fewer stages than impactors so resolution of particle size is more limited. A four-stage liquid impinger is one example (Snell and Ganderton 1999). The two-stage liquid impinger is convenient for rapid analysis of the fine particle fraction or "respirable dose" of an aerosol (LeBelle 1997). The fine particle fraction is defined differently by various sources, but in general, it is the fraction of particles in an aerosol of a size small enough to penetrate the lungs. The two-stage liquid impinger separates particles into two size ranges, those greater than and those less than 6.4 μm (Pezin et al. 1996).

## Filtration

Filtration may be used either for particle size determination or for particle segregation for other purposes (Hickey 1992). Use of a graded sequence of filters in cascade provides particle sizing data much like sieving. The pressure drop at each filter stage must be monitored to ensure that clogging does not occur anywhere in the cascade. Filters are also useful for the collection of aerosol particles for microscopy. Membrane filters are particularly useful for microscopy because their structure, consisting of tortuous channels, tends to collect particles on their upper surface (Allen 1968). The filter can then be used directly for microscopy. Filters operate on the principle of interception and impaction. Large particles impact directly on filter fibers. Smaller particles may avoid initial impact but are intercepted as they leave the streamlines of air passing around the filter fibers.

## Andreasen Pipette

The Andreasen pipette instrument uses a sedimentation technique for particle sizing. As discussed in Chapter 5, the settling velocity of particles in a fluid depends upon their diameter. The Andreasen pipette takes advantage of this effect by measuring concentration changes at the bottom of a vessel as suspended particles settle in the dispersing liquid. The sedimentation vessel is filled with a powder sample dispersed in liquid to a height of 20 cm. In order to avoid aggregation effects in the liquid, the sample concentration should be 2 percent or less by volume (Martin 1993). The powder sample is dispersed by shaking, or preferably, by slowly turning the vessel upside down and back for a number of repetitions. Sampling should not begin for one minute, to allow particles that may have had an initial velocity to reach their settling terminal velocity. A 10 mL sample is then drawn from the pipette, generally at 20 second intervals. The sample is then drained, dried, and weighed or otherwise analyzed for concentration. Particles collected by the tube at height h are between the sizes that settle in the time between $t_i$ and $t_{i+1}$, where the times are for the ith and (i+1)th sample. This particle diameter, $d_p$, is derived from Stokes' equation and is

$$\sqrt{\frac{18h\eta}{(\rho_p - \rho_f)gt_{i+1}}} \le d_p \le \sqrt{\frac{18h\eta}{(\rho_p - \rho_f)gt_i}} \qquad (6.6)$$

where $\eta$ is the kinematic viscosity; g is the gravitational constant; and $\rho_p$ and $\rho_f$ are the densities of the particle and the fluid, respectively.

The Andreasen pipette method is inexpensive. However, it does have drawbacks: Potential for error occurs because the sampling tube interferes with sedimentation, especially directly below the tube, and because the tube samples a spherical rather than a planar region. The sedimentation fluid must be kept at constant temperature for the duration of the experiment because viscosity is a function of temperature. In spite of these limitations, the method can still be accurate to within 2 to 5 percent

## Virtual Impinger

The virtual impinger was developed to minimize the bounce and reentrainment problems of the cascade impactor (Kaye 1981). In this impinger, no collection surfaces are present. Instead, an airstream is directed toward a container with a volume of static air. The airstream is diverted around the sides of the vessel because air cannot be conducted through it. This static air forms a virtual impaction surface. Larger particles will be carried by their inertia out of the streamline and into the vessel, where they are collected. Virtual impingers operate with less turbulence at the vessel edges if there is a slow flow of air from the vessel. One example is the dichotomous sampler in which 1/49th of the total flow passes through the collection vessel. A collection filter is placed in the vessel to capture particles entering the opening. Virtual impingers can be placed in series with an increasing airflow rate moving down the series to produce a particle size distribution measurement.

## Centrifugation

Centrifugation works on the same principles as sedimentation, except that the gravitational sedimenting force is replaced by a centrifugal force generated by a spinning sedimentation instrument. Forces much greater than gravity can be generated with the centrifuge to increase settling and reduce experimental time. The gravitational term in Stokes' equation [see equation (5.35)] for particle terminal velocity is replaced by a centrifugal term, $a_r$, as:

$$v = \frac{2a_r d^2(\rho_p - \rho_f)}{9\eta} = \frac{2\omega^2 r d^2(\rho_p - \rho_f)}{9\eta} \tag{6.7}$$

where ω is the radial velocity and r is the radial position of the particle relative to the center of rotation. Since particle velocity is a function of radial position as v = dr / dt, equation (6.7) can be rearranged to yield:

$$\frac{dr}{r} = \frac{2\omega^2 d^2(\rho_p - \rho_f)}{9\eta} dt \tag{6.8}$$

Integrating equation (6.8) and solving for particle diameter d yields the diameter of a particle with initial position $r_1$ and position $r_2$ at a time t:

$$d = \sqrt{\frac{9\eta \ln(r_2 / r_1)}{2\omega^2(\rho_d - \rho_f)t}} \tag{6.9}$$

This diameter is used as the diameter of particles sampled from the centrifuge at time t.

In operation, the centrifuge is brought to the chosen speed and the spin fluid introduced from a coaxial injection needle. Once the fluid has stabilized, the powder sample dispersed in the spin fluid is injected into the centrifuge. Samples are taken at time intervals, concentrations measured, and diameters calculated from equation (6.9). An advantage of this approach over gravitational sedimentation is the accuracy with which a sample can be introduced. The sample is introduced with zero initial velocity and without turbulence. The density and viscosity of the spin fluid can be chosen to achieve reasonable centrifuging times. Centrifugation gives results accurate to less than 2 percent uncertainty in the cumulative weight undersize distribution. As with gravitational sedimentation, the sample concentration should be less than 2 percent to avoid hindrance of sedimentation. The particle density must be known in order to calculate diameters. Errors in its measurement will contribute significantly to errors in the results. Ultracentrifuges that spin at speeds of up to 60,000 revolutions per minute (rpm) can be used to measure particle size distributions for particles as small as 0.01 μm.

## Approaches to Calibration

Closely related to the measurement of particle size, distribution, and morphology is the need to calibrate the instruments performing these tasks. For direct imaging techniques, calibration is straightforward. A particle of known diameter is required, or precisely manufactured graticules or micrometer stages may suffice. Microscopy calibration standards include manufactured spheres and naturally occurring particles. Manufactured standards are constructed from glass, quartz, and polystyrene. These standards may be purchased in sizes ranging from 0.4 to hundreds of microns. Pollen grains are nearly monodisperse and can be used for calibration. Pollen grains range in size from approximately 15 to 80 μm.

Calibration techniques for other methods of particle sizing are more complex and may require specialized equipment. For these instruments, the physical and chemical properties of standards may need to be precisely controlled. Particle generation for calibration of aerosol particle sizing equipment requires continuous generation of particles over the entire sampling time. They must further provide an aerosol of sufficient concentration for measurement by the instrument.

### Vibrating Orifice Monodisperse Aerosol Generator

The vibrating orifice monodisperse aerosol generator (VOAG) creates particles via the breakup of a jet of liquid forced through an orifice that is driven by a periodic vibration. Vibration of the orifice is provided by a piezoelectric ceramic. This vibration breaks up the jet into uniform droplets. The resulting stream of particles is diluted with heated air to evaporate the solvent and to minimize droplet coagulation.

The liquid feed rate, Q, and the frequency of vibration, f, are the primary determinants of the generated particle diameter with

$$d = \left( \frac{cQ}{\pi f \rho_p} \right)^{\frac{1}{3}} \tag{6.10}$$

where c is the mass concentration by volume of solute. Particle diameters ranging from 0.5 to 50 μm can be achieved (Berglund and Liu 1973). Particle size distributions are very monodisperse with $\sigma_g < 1.05$. Larger particle production is not practical because evaporation time is lengthy for production of solid particles and concentrations are low. Typical particle concentrations generated with a 10μm orifice are around $10^3$ $cm^{-3}$. Concentration decreases with decreasing orifice size because the vibration frequency also decreases (and, thus, the number of particles generated per second decreases).

The liquid fed through the orifice should be filtered. However, orifice plugging can still be a problem. For solid particles where evaporation is required, appropriate choices in solvent, concentration, and compound are important. Evaporation rates that are too rapid will result in the production of less solid and rough-surfaced particles.

## Spinning Disk Generators

The spinning disk method of monodisperse particle generation produces less concentrated aerosols (around 100 $cm^{-3}$) than the VOAG instrument. It does have the advantage that, without an orifice, plugging is not an issue, so operation is very reliable. The spinning disk is a rotor-stator system in which high-pressure air is directed over vanes on the rotor, causing it to spin rapidly, up to 70,000 rpm. A liquid feed is directed onto the top center of the disk, so droplets are flung outward by centrifugal force. Typically, larger primary droplets and smaller satellite droplets are generated. The satellite droplets are deflected by the airstream and rejected.

Droplet sizes ranging from 10 to 150 μm are typical. The size is a function of the rotation angular velocity ω, liquid concentration c, rotor diameter D, and liquid surface tension T, according to the equation:

$$d = \frac{K}{\omega}\left(\frac{T}{D\rho_l}\right)^{\frac{1}{2}}\left(\frac{c}{\rho_p}\right)^{\frac{1}{3}} \tag{6.11}$$

where the constant K is an empirical term ranging from 2.3 to 7 for water (Byron and Hickey 1987). The densities $\rho_l$ and $\rho_p$ refer to the liquid and the particle, respectively.

As stated, suspensions can be used with the spinning disk more readily than with the VOAG because the disk will not become plugged. However, the cautionary statements concerning particle morphology as a function of evaporation rate apply for this instrument as well.

### Charge Neutralization

Generation of monodisperse particles by charge neutralization relies on the balance of electrostatic and aerodynamic forces on particles. In operation, a polydisperse aerosol is passed through a charging chamber containing a radioactive source (usually $Kr^{85}$). Particles receive a single positive, negative, or zero charge. The particles are then passed to a differential mobility analyzer that has a grounded sheath surrounding a central, negatively charged rod. Negatively charged particles are attracted to the outer sheath, positively charged particles are attracted to the central rod, and uncharged particles pass with the airflow and are pumped out of the chamber. Particles of an appropriate diameter will move toward the central rod but will just miss the rod's end and instead be pulled into a tube below the rod. By adjusting the flow rate and voltage of the central rod, the aerodynamic and electrostatic forces on particles can be balanced to "dial in" the desired monodisperse particle size output.

## Conclusions

This chapter discusses a number of particle sizing techniques ranging from direct and indirect imaging of particles to physical methods of separation by size. The choice of method depends on the use of the powder, its physical properties, and its expected size and distribution. As with any process, the cost of measurement, both monetary and in time, is a consideration. The cost of measurement also includes the quantity of powder required by the chosen method for determining the particle size and whether the method is destructive to the sample. This concern will probably be overriding when sizing expensive biotechnology products.

The sizing method employed should be appropriate to the use of powder being measured. For example, measurement of powders intended for dispersion as aerosols should be measured in their dispersed state. Further, the mechanism used to disperse the powder in actual use should be the dispersion method employed for the measurement, if possible. In this example, although geometric diameter may be measured to aid in quantification of the powder, some means of measuring the aerodynamic diameter of the powder should also be utilized. The aerodynamic diameter is the measure of greatest interest because it will determine the distribution of the powder in the lungs when inhaled as an aerosol. As discussed in Chapter 4, the choice of appropriate statistics is also important. For the pharmaceutical aerosol example, the mass median aerodynamic diameter is the most important measure of size because it is the particle mass that will determine its therapeutic effect.

Also discussed in this chapter is the need to employ more than one particle sizing technique to construct a more complete picture of the product. One of the methods employed should almost invariably be some form of direct imaging. Use of direct imaging serves as a reality check and will ensure that any erroneous data produced by other methods will not go undetected. Many of the more modern techniques of particle sizing, such as laser diffraction, tend to be treated as a black box with unquestioned results. Use of a confirmatory technique is extremely important—at a minimum initially—to validate the instrument's operation.

# 7

# Methods of Particle Size Measurement and Their Importance for Specific Applications (Instrument Synergy)

This chapter draws together, in a short essay, the topics of particle size, particle size statistics, instrumental analyses, and pharmaceutical application (but not biological implication). Each of these is explored in detail in the preceding chapters. The interplay among these factors and their influence on the desired pharmaceutical application is explored in this chapter. For example, the diameter measured (projected area, surface area, or volume) and the statistics employed (by number or by mass) should be dictated by the intended application. These considerations will, in turn, usually suggest an appropriate measurement instrument or instruments.

## An Essay on Measures of Diameters

The range of options for defining individual diameters of particles is described in Chapter 4. Particle diameters can be expressed for a number of characteristics of the particle. These properties include projected area, surface area, volume, and mass. Each of the properties inherent to the particle is an accurate description of the particle, but in any given pharmaceutical process, a particular property will often be the most relevant for predicting the particle's behavior. In some instances, the particle size determined by one method may not be predictive of the particle's performance if the appropriate measurement technique is not chosen. Brittain et al. (1991) gave the example of cohesive powders whose flow properties indicated a larger particle diameter than measured by laser diffraction because the particles were dispersed in a liquid before measurement.

Diameters determined by optical imaging are one means of describing the primary particle size. Optical methods allow a determination of the particle size and morphology. As an example of the relevance to pharmaceutical processing, the packing and flow properties of a powder are influenced by its morphology. Spherical particles will pack and flow more efficiently than highly irregularly shaped powders. Efficacy of tablet production, then, may best be predicted by diameters that depend upon particle morphology. The ability to increase density of powders during compaction is the basis for tablet production and is a function of diameter (Mbali-Pemba and Chulia 1995). Smaller original particle sizes will result in compacts with increased tensile strength (Alderborn 1996). The powder flow properties will be important for uniform filling of tablet dies (Schmidt and Rubensdorfer 1994). Packing properties will affect tablet hardness and, subsequently, dissolution properties (Menon, Drefko, and Chakrabarti 1996). Particle morphology is relevant to aerosol dispersion because the flow properties of a powder are determinates of the effective dispersion of powder aerosols from dry powder inhalers (Concessio 1997).

This discussion should make it clear that characterization of powders for tableting or aerosol dispersion should include some optical measure of the physical diameter and morphology of the powder particles. However, it should also be noted that these measures alone will not be sufficient for optimizing either process. In the case of tableting, dissolution properties will also be a function of the surface area of the powder. For aerosol dispersion, a measure of the aerodynamic diameter is necessary for prediction of the deposition properties in the lung.

The primary particle size will determine which instrument can be used to provide optical particle size data. Conventional light microscopy may be sufficient for powders intended for tableting, whereas electron microscopy must be used for inhalation aerosol powders because their diameter must be below 5 μm for lung deposition. Although the aerodynamic diameter is the most relevant measure for aerosol dispersion and deposition, this measure alone is not sufficient for prediction of therapeutic effect. The aerodynamic diameter is related to the geometric diameter through the particle density. Porous particles have recently attracted attention for pulmonary deposition. The particles are engineered to be less than unit density by virtue of a porous structure with trapped air volume within the particle. Because these particles have a mass density significantly lower than unity, particles with an aerodynamic diameter (diameter of a unit density sphere with the same settling velocity as the particle in question) in the respirable size range can be achieved despite having a geometric particle size greater than 10 μm (Edwards et al. 1997). However, the therapeutic effect of the drug particle is determined by the mass of the particle deposited. Therefore, when characterizing aerosol size, a measure of the aerodynamic diameter must also take into account the particle density.

The surface area diameter is another example of a measure that is predictive of powder performance in a desired process. Particle surface area has implications for particle flow, aggregation, and dissolution behavior (Hickey 1992). As stated previously, the disintegration and dissolution properties of a tablet depend on the surface area of the powder in the tablet (Menon, Drefko, and Chakrabarti 1996). The dissolving medium may be able to penetrate pores in the particles, if they exist, aiding in disintegration and dissolution. Thus, measurement of the surface area will be predictive of dissolution rate. Tablet compact hardness depends on the particle and bonding surface areas of the constitutive powder. These surface areas are functions of the particle size and shape (including internal surface area of the particles due to their effects on fragmentation) (Nyström and Karehill 1996). Other studies support the conclusion that compaction properties of the powder in the tablet depend upon the surface area (Rime et al. 1997). For particles of very unusual morphology, the interparticulate contact area, measured by nitrogen adsorption, determines the compact hardness of the powder. For the particular powder described in this study, the packing properties could not be ascribed to any other material-related property of the powder.

The volume diameter has relevance to pharmaceutical technology, especially in regard to delivery of particles in an aerosol. The volume diameter can be measured by electrical resistance techniques, as described in Chapter 6. As previously described, the therapeutic effect generated by a deposited aerosol particle is determined by the deposited mass, directly related to the volume diameter. Care must be taken, however, for particles with trapped air volume, because the volume diameter will be related to the mass through the particle density. The mass dependence on volume scales with the cube of the diameter, so changes in diameter will have a significant effect on the deposited mass.

In most cases, more than one measure of particle size and characteristics must be collected to predict performance, even if one measure of performance is taken. Foster and Leatherman (1995) discussed how multiple measures are needed to describe performance for particles created in a complex process such as spray drying. In their study, the morphology and flowability of the spray-dried powders remained the same across multiple equipment parameters during scale-up. Overall performance, however, changed significantly due to changes in particle size and distribution and bulk density.

## Conclusion

This short chapter is intended to tie together previous chapters that described particle size measures and their instrumentation. It is intended to show that measures of particle size cannot be thought of independently from the means of measurement of particle size. The next chapter takes these ideas a step further and ties together particle size and performance in pharmaceutical processes.

# 8

# Particle Size and Physical Behavior of a Powder

This chapter discusses the correlation between particle size and the physical behavior of a powder, with particular emphasis on pharmaceutical processes. Pharmaceutical processes involving powders encompass a variety of dynamic effects ranging from powder flow and mixing, to dispersion as aerosols or liquid suspensions, to granulation and compression. The performance of the powders involved in pharmaceutical operations is highly dependent upon the particle size and particle size distribution of particles in the powder. This chapter discusses these effects in detail and provides examples of how pharmaceutical processes are impacted by particle size.

## Selection of the Appropriate Particle Size Expression

Chapter 7 provided an essay on the importance of instrument synergy, that is, the consideration of the appropriate size descriptor for the intended application of the particles in question. It cannot be emphasized strongly enough that the particle size and statistics used to describe the powder should be chosen according to the pharmaceutical process in which the powder is to be used. A number of examples have been given relative to measures employed for the various intended uses. Although this concept is important for static effects, such as dissolution, it is even more critical with respect to particle dynamics. The dynamics of particles are highly dependent upon particle size; in many cases, the physics of the particle may vary with the square or cube of the particle diameter. For example, the terminal settling velocity of a particle in a viscous medium (air or water) increases with the square of the particle's geometric diameter. For particles intended for inhalation (i.e., less than 10 μm in size), the Cunningham slip correction factor must be added to the particle dynamics to account for the effect of ambient gas molecules with mean free path motion on the order of the size of the particles themselves. The Cunningham slip factor alters the settling velocity of particles by a factor of 1.02 for 10 μm particles, by 1.17 for 1 μm particles, and by 2.97 for 0.1 μm particles (Crowder et al. 2002). Therefore, it should be obvious that particle size is of critical importance for inhalation powders, with differences of even a few microns significantly altering the dynamics of the particles. As can be seen from this example, no description of inhalation powders would be complete without stating the geometric diameter of the particles.

Although the geometric diameter of powder particles is important for most processes, other factors may be more influential in powder performance. For example, powder compressibility is determined primarily by particle volume, and dissolution is determined by particle surface area. These properties are discussed in detail in this chapter.

## Powder Flow and Mixing

The processing of most pharmaceutical products that incorporate powders will have a step that involves vibration. This step will often be a transfer process, such as die filling for tablet production or blister packing for aerosol unit dose production. Likewise, mixing processes are widely used in the preparation of pharmaceutical products. Tablet and dry powder aerosol dosage forms require mixing the active particles with excipients. Both of these processes are influenced by particle size. Milling, wherein particle size is reduced, is another example of a size-dependent process. Each of these processes are examined in more detail in the remainder of this chapter.

## Vibration and Powder Transfer

Vertically vibrated powders have provoked significant interest among physicists. Vibration— in particular, sonic vibration—historically has been used in industrial powder processing. In a relevant industrial application, gas fluidized beds were simultaneously vibrated using sonic energy generated by a loudspeaker (Leu, Li, and Chen 1997; Mori et al. 1991). The gas velocity required to fluidize the bed decreased when sonic energy was added to the system. The method was efficacious even in beds with particle sizes below 1 μm.

Physically vibrated powders can form heaps or complex patterns. The powder bed can be made to dilate if the vibration conditions are correct (Van Doorn and Behringer 1997). For vertically vibrated powders, the amplitude of vibration is generally given by the function $z = a\cos(\omega t)$, where $\omega$ is the angular frequency in radians/second and a is the amplitude. For convection in the powder bed to occur, the vibrational acceleration must be larger than the gravitational acceleration. This condition can be written by using a dimensionless acceleration parameter. Dilation occurs when

$$\Gamma = \frac{a\omega^2}{g} > 1 \tag{8.1}$$

where g is the gravitational constant. The dilation effect was found to be dependent upon the ambient gas. Dilation is significantly reduced if the ambient gas pressure drops below about 10 Torr (Pak, Van Doorn, and Behringer 1995). The postulate that interstitial gas is required for powder bed dilation was confirmed by experiments in a chamber where the interstitial gas was allowed to escape through holes in the side of the chamber as the powder was vibrated.

At vertical vibration accelerations with Γ greater than approximately 2.4, patterns appear on the surface of the granular material. Stripes, squares, or localized oscillations have been observed (Jaeger, Nagel, and Behringer 1996; Melo, Umbanhowar, and Swinney

1995; Pak and Behringer 1993). A study by Duran (2000) demonstrated that powder beds tapped from below exhibit ripples and that particles can be ejected from the powder surface by tapping.

Vibration of powders causes segregation by size. Horizontally vibrated powders in closed containers exhibit a surprising phenomenon wherein large particles move to the bottom of the powder bed (Liffman, Metcalfe, and Cleary 1997). This contrasts with the normally observed percolation that causes small particles to move to the bottom of the bed (Barker and Grimson 1990). This effect was determined to be due to upward convection near the walls of the vibrating container with a subsequent downward convection in the center.

Vibration can also result in consolidation of the powder bed. For consolidation, the degree of densification is a size-dependent phenomenon. Packing density varies with particle size when the powders have been sieved into discrete size ranges. Vertical vibration leads to denser packing of the powder bed than does horizontal vibration (Woodhead, Chapman, and Newton 1983). Packing due to vibration is frequency dependent, but the frequency is largely particle size independent. These authors found that the optimal frequency of vibration for consolidation of lactose powders was between 100 and 200 Hz. These powders were divided into eight size ranges from around 20 to 150 μm.

Size segregation can also occur for nonvibratory powder transfer. In chute flow, fingers of powder appear across the advancing powder front due to the size segregation mechanism (Pouliquen 1997). In this situation, the coarser particles rise at the free surface due to the same percolation mechanism discussed previously. This leads to a recirculation zone at the front. Filling of a hopper leads to segregation of particles by size, with smaller particles tending toward the center of the hopper (Standish 1985; Schulze 1994). Some remixing of particles can occur upon emptying of the hopper, but a net segregation is still observed. The degree of segregation depends on the manner in which the powder is discharged. The discharge flow pattern in turn depends upon the hopper geometry, orifice size, and wall roughness (Kafui and Thornton 1997; Khati and Shahrour 1997). Hoppers or silos can empty by mass flow by which the entire content moves through the outlet or funnel flow in which only the center of the bulk powder flows through the outlet, leaving a dead zone (Schulze 1994).

Filling properties of capsules by a nozzle system can be similar to those by hopper flow. Specifically, the powder-wall interaction in the nozzle has been found to impact capsule filling in the same manner as for flow from a hopper (Jolliffe, Newton, and Walters 1980). The particle size of the powder used in the fill is an important variable in capsule-filling properties (Jolliffe and Newton 1982; Hogan et al. 1996). When these methods of powder transfer are used, the existence of segregation mechanisms must be considered in the evaluation of end-product properties.

Pneumatic conveying may be used to move bulk solids over large distances. The process may reduce particle size due to the large amount of energy required to move the particles. In addition, segregation may occur due to particle size dependence in the deaeration behavior (Hilgraf 1994).

In general, particle size descriptors relevant to powder-handling processes are the geometric diameter and volume diameter. Both of these diameters determine the interparticulate forces and contact areas between powder particles. These diameters are also relevant to the bulk and tap densities, which, in turn, determine the settling properties that affect powder flow and transfer. For example, a direct correlation has been demonstrated between bulk density and capsule fill weight (Newton and Bader 1981).

## Mixing and Milling

The process of mixing finds many applications in pharmaceutical processing. Much of the knowledge of the process is gained by experience (Poux et al. 1991), and the fundamental physics of the process is little understood. In addition, numerous types of mixers are used in pharmaceutical processing, from rotating vessels to vessels with moving baffles, blades, or screws (ribbon blender). Each mixer will have different performance properties. Mixing cannot be viewed in isolation because the filling or emptying of the mixer can result in segregation, as discussed in the previous section.

A number of generic studies of mixing of granular materials can be found in the physics literature. One example examined the mixing of granular materials in a tumbling mixer. In that study, Khakhar et al. (1999) developed a generic model for an arbitrarily shaped convex mixer (any two points in the mixer can be connected by a line that does not bisect the boundary of the mixer). The model conditions required that the rotation rate be sufficient for particle transport to be continuous, rather than consisting of discrete avalanches, and that flow be constrained to an upper surface layer of limited depth. These conditions have been observed experimentally (Henein, Brimacombe, and Watkinson 1983; Nakagawa 1997). Under these conditions, a layer-by-layer map of particle trajectories was generated. For circular mixers, the trajectories were almost axially symmetric, and mixing between axial layers was poor. If the mixer geometry was elliptical or square, the trajectories became chaotic. When the model conditions were initiated with two segregated types of particles, the degree of segregation for a circular mixer decreased by only 10 percent after 20 rotations versus 35 percent for a square mixer. Khakhar et al. (1999) performed experimental simulations using 0.8-mm glass beads that provided results confirming the computational model. As they noted, particle sizes in their model were the same, whereas under practical mixing conditions, particles exist in a range of sizes. In this case, demixing by segregation of particle sizes would counter mixing by chaotic advection (Khakhar et al. 1999). The authors suggest that more complex blenders, such as V-blenders, probably perform by harnessing chaotic advection. By elucidating a model of mixing, Khakhar et al. (1999) have provided a tool that could be employed in the design of efficient mixers.

A purely geometrical model has confirmed the time scale required for mixing in a circular drum (Dorogovtsev 1997). This model calculated the number of rotations required for the mixing of powder layers stacked vertically in the mixer. Here, mixing was defined as the occurrence of sufficient avalanches for the interface between powder layers to reach the surface through rotation. Uniform mixing, then, was not implied, but simply the disappearance of the initially pure surface regimes of the top powder. More than 10 revolutions were required for a drum 5 percent filled with greater than 25 percent of the volume occupied by the top powder. The mixing time decreased as the volume fraction of the top powder decreased below 25 percent and as the filling of the drum approached half full. For differently colored sodium chloride crystals that were otherwise identical particles, different authors found that the optimal circular mixer fill was 23 percent (Metcalfe et al. 1995).

Mixing can induce size reduction by attrition in coincident milling. Alternatively, cohesive materials may aggregate during mixing due to surface tension forces that, in turn, are due to liquid bridging, mechanical interlocking, or electrostatic forces (Bridgwater 1994).

Ball milling is an example of another highly complex system for which knowledge of the relationships between the inputs (milling time, powder load, size and quantity of milling media, and rotation speed) and the output (milling rate, particle size, and distribution) is largely based on experiential observation. Caravati et al. (1999) modeled the motion of a single ball in a ball-milling device. Examination of the ball motion, particularly the velocity of the ball before impacts, could lead to predictions of the milling rate. The model was based upon a SPEX mixer/mill (model 8000, SPEX CertiPrep, Metuchen, N.J.). An experimental determination of the motion of the vial and the frequency of collisions of the ball and vial using a piezoelectric shock sensor to detect the ball validated the numerical simulation. The time series of impacts of the ball depended upon the powder charge in the mill. For a large powder charge, collisions of the ball and the wall of the mixer were plastic due to the cushioning effect of the powder between them. With plastic collisions, the impacts were highly periodic. As the powder charge was decreased, the collisions became more elastic. As a result, the ball motion lost its periodicity and eventually became aperiodic. A strange attractor was observed when the ball position and velocity were plotted in phase space for the plastic collisions. For this case, determination of the Lyapunov exponents confirmed that the motion of the ball was hyperchaotic. This study (Caravati et al. 1999) was the first step in the development of a means to predict milling properties from a physical model of the milling system. Correlation of the degree of chaos with the mill output parameters could be applied to optimization of input parameters for the desired output.

## Dispersion

Powders have been dispersed as aerosols for delivery to the lungs or airways since the invention of the modern dry powder inhaler (DPI) in the 1960s (Clark 1995; Hickey and Dunbar 1997). Interest in this means of drug delivery is accelerating due to environmental concerns around the pressurized metered dose inhaler (pMDI), the device that delivers the majority of respiratory drugs, and due to interest in delivering large molecules for systemic therapy (Crowder et al. 2001). The discussion in this section is limited to a brief summary of particle size effects in therapeutic aerosol dispersion. Particle size is of critical importance because it affects not only dispersion but also deposition in the lungs. The range of particle sizes that can deposit in the lungs is limited to approximately 1 to 5 μm. Entire texts are available that discuss the physical pharmacy of aerosol dosage forms (Hickey 1992) or biological implications (Hickey 1996) of aerosol drug delivery.

At rest, granular materials act as a plastic solid (Castellanos et al. 1999). In this state, the interparticle spacing is small, particle velocities are near zero, and the stresses on the powder bed are independent of velocity. The properties of the plastic regime determine the stability and slopes of heaps of material. Flow begins when a granular bed makes a transition to an inertial regime (Evesque 1991). Two other regimes of collective behavior exist. The fluidization regime is characterized by interparticle spacing on the order of the particle size. The interstitial fluid is the agent of momentum transfer, and thus must overcome interparticle forces to initiate fluidization. Finally, the fourth regime is that of entrainment or suspension of the granular particles. Here the interparticle spacing is very large, and interaction between particles is negligible. The velocity of the particles is close to the velocity of the entrainment fluid. Generation of a powder aerosol involves the transition through the four states with the dynamics of each step being critically dependent upon the particle size. The conversion of a static powder bed to a powder aerosol requires dilation, flow, fluidization, and finally aerosol production (Dunbar, Hickey, and Holzner 1998).

The efficiency of dispersion must also be considered in the design and formulation of dry powder aerosols. Dispersion requires the powder to overcome the interparticulate forces binding particles in bulk powder and to become entrained as single particles in the inhaled airstream. For respirable particles, these forces are dominated by the van der Waals force. All other factors being equal, the van der Waals force will decrease as the geometric particle size increases. Thus, in general terms, the probability of deposition in the deep lung acts in opposition to the efficiency of dispersion for DPIs (Hickey et al. 1994).

The traditional approach to overcoming the difficulties in dispersion of particles suitable for deposition in the lungs has been to blend these particles with larger carrier particles of lactose. The blended system optimizes interparticulate forces to create a mixture that will respond to the fluidization process. This system requires that the smaller particles be deaggregated from the larger carrier particles before they can enter the airways (Hickey et al. 1994). In currently approved DPI devices, the forces required for deaggregation are provided by the patient, who generates shear forces in the DPI during inhalation (Crowder et al. 2001). Blending of ternary mixtures has been shown to increase the proportion of respirable particles delivered by some DPIs (Zeng, Pandhal, and Martin 2000). These blends comprise the micronized drug, the lactose carrier, and fine particles, typically lactose. The fine particles establish competition for binding sites and can result in a higher proportion of drug particles binding to weaker sites.

Engineering of low-density particles has gained popularity in recent years as an approach to overcoming the opposition between efficient dispersion and efficient aerosol delivery to the lungs (Edwards, Ben-Jebria, and Langer 1998; Edwards et al. 1997). This approach takes advantage of the relation between aerodynamic diameter, particle density, and geometric size. The balance of gravitational settling forces and the aerodynamic drag force determines this relationship. When these forces balance, the particle is said to have reached its terminal settling velocity. The aerodynamic diameter is defined as the diameter of a unit-density particle that has the same settling velocity as the non–unit-density particle being analyzed (Reist 1993a). Thus, the equation relating the geometric diameter to the aerodynamic diameter for spherical particles is given by

$$d_a = \sqrt{\rho} d \tag{8.2}$$

where $d_a$ is the aerodynamic diameter; $\rho$ is the (non-unit) density of the particle; and d is the particle's geometric diameter. See Chapter 6, especially equations (6.4) and (6.5), for further details. Aerosols composed of particles with equal aerodynamic diameters but differing densities will necessarily have differing geometric diameters. Clearly, particles of less than unit density will have aerodynamic diameters that are smaller than their geometric diameters. Therefore, a particle with a geometric diameter greater than 5 μm can be "respirable" provided the density is sufficiently small. This property, which allows large particles to penetrate deep into the lungs, is employed in the manufacture of large porous particles.

## Granulation and Compression

Granulation is the production of a powder by aggregation of smaller particles, typically used for the production of tablets and capsules. Bonding mechanisms employed during granulation are adhesion due to capillary pressure, adhesion due to liquid bridges, and solid bridges formed during drying (Ormós 1994). The liquid binding forces are responsible for the resultant particle size, whereas solid bridging is primarily responsible for granulate strength. Wet granulation involves the wetting of the seed particles by a liquid to create controlled conditions of agglomeration. The powder bed is kept in motion during this process either by flow through gas or other fluidization mechanisms such as vibration. Particle size effects during vibration were discussed in the previous section. The important point is that the energy input for fluidization must be increased during the process to overcome gravitational forces as the particles increase in size.

Compression of powders involves four stages: slippage and rearrangement of particles, elastic deformation of particles, plastic deformation or brittle fracture of particles, and compression of the solid crystal lattice (Doelker 1994). These mechanisms may not occur sequentially; for example, smaller particles formed by fragmentation may undergo another rearrangement step.

Tableting requires control of the particle size distribution of the components because this affects the final homogeneity of the active particles in the tablet and their bioavailability (Lewis and Simpkin 1994). In particular, matching of the active and excipient particle sizes typically leads to better homogeneity due to a reduced tendency for particles to segregate. Much of this effect arises due to the flow properties of the powders during blending steps. The dissolution properties of the active properties depend upon the particle surface area , and a drug with a poor aqueous solubility requires a larger surface area than a more soluble drug. In addition, greater tablet strength is often achieved in tablets formed of fine particles (Rudnic and Kottke 1996). Clearly, determination of the particle surface area is important for powders used in tableting. Other problems arising from poor control of particle size distribution include variation in tablet weight and hardness. Variations in flow properties due to particle size can result in poor flow from hoppers and bins or flooding of tablet machines.

## Conclusion

This chapter ties the specific utilization of pharmaceutical powders to the appropriate measure of particle size. The physical behavior of a powder particle is closely correlated to at least one measure of its size or characteristics. One or two of these characteristics will frequently be more predictive of its performance than others in a given pharmaceutical process. However, the choice of characteristic highly depends upon the desired process.

The next chapter discusses the biological implications of pharmaceutical particle size. Ultimately, this measure must be the most important concern to the pharmaceutical scientist.

# 9

# Clinical Effect of Pharmaceutical Particulate Systems

Drug delivery systems aim to provide accurate and convenient drug administration, to produce the desired therapeutic response with minimal side effects and to maximize patient compliance (Robinson, Narducci, and Ueda 1996). The majority of drug delivery systems achieve the required drug levels at the site of action by attaining adequate blood levels in the general circulation. Rapid and complete drug absorption into the bloodstream is desired for the following reasons (Mayersohn 1996): Greater response is achieved with higher blood levels in cases for which a relationship between circulating drug concentration and therapeutic response has been established. More rapid therapeutic response is achieved with faster absorption. Pharmacological response is uniform and reproducible with rapid and complete drug absorption. Less chance of drug degradation or interactions occurs with rapid drug absorption.

Bioavailability is a measure of the rate and amount of unchanged drug that reaches the systemic circulation following administration of the dosage form. Drug bioavailability and clinical response is primarily affected by the physicochemical properties of the drug, physiological factors, and dosage form variables. Manufacturing variables and environmental factors may also affect drug bioavailability. The clinical response is a complex function of the interaction of these variables (Mayersohn 1996). Physicochemical properties of a drug may include aqueous solubility, $pK_a$, pH-solubility profile, lipid-water partition coefficient (log P), particle size and size distribution, crystalline and polymorphic forms, solvation and hydration state, salt form, molecular weight, and stability in the solid and solution states. Physiological factors may include pH, temperature, surface area, surface tension, volume and composition of biological fluid, disease state, and gender and age of patient. Dosage form and manufacturing variables include the intended route of administration, presence of excipients, and manufacturing methods. Environmental factors include temperature and relative humidity.

Common routes of drug administration include oral, sublingual, buccal, parenteral, respiratory, nasal, transdermal, rectal, vaginal, ocular, and otic. The most appropriate route of administration for a drug depends on the desired target site of action and may be limited by the physicochemical and pharmacokinetic properties of the drug, desired onset and duration of action, and convenience or status of the patient (Robinson, Narducci, and Ueda 1996). However, the most convenient route of administration is frequently used because targeted delivery may be impractical or impossible or the site of action is unknown (Oie and Benet 1996). Differences in the intensity and duration of therapeutic response are obtained from drug administration by different routes.

Targeted drug delivery to the site of action maximizes the pharmacological effect and minimizes side effects relating to unwanted responses at other sites. Targeted delivery provides more rapid onset of action because it avoids the process of distribution, and it reduces the amount of drug required because the drug is not diluted or eliminated en route. The drug must remain at the site of application for a sufficient time to enable penetration through the membrane to the site of action for local action (Oie and Benet 1996).

## Oral Delivery

The oral route is the most commonly used route for systemic drug delivery. It is both convenient and economical (Robinson et al. 1996). However, some drugs are chemically or enzymatically degraded in the gastrointestinal tract (GIT). Drug absorption following oral administration may be limited by drug instability and poor intrinsic membrane permeability, resulting in inefficient and erratic drug therapy (Mayersohn 1996).

The GIT represents an important barrier and interface with the environment; its primary functions involve the processes of secretion, digestion, and absorption. Vomiting and diarrhea are the primary defense mechanisms employed by the GIT to void noxious or irritating materials (Mayersohn 1996). The main sites in the GIT are the stomach and the small and large intestines. The liver, gallbladder, and pancreas are important organs involved with the digestive and absorptive functions. The functions of the stomach are storage, grinding, and mixing. The stomach contents are emptied into the duodenum in a controlled manner. The small intestine comprises the duodenum, jejunum, and ileum, with average lengths of 0.3, 2.4, and 3.6 m, respectively (Mayersohn 1996). The surface area of the small intestine is 100 $m^2$ (Chien 1992b). The large surface area of the small intestine is due to the folds of Kerckring, the villi, and the microvilli. The main functions of the small intestine are digestion and absorption. The functions of the large intestine are the absorption of water and electrolytes and the storage and elimination of fecal material (Mayersohn 1996).

The entire GIT is highly vascularized, receiving about 28 percent of the cardiac output (Mayersohn 1996). The rapid blood perfusion at the site of absorption represents a "sink" for drug compounds, with the concentration gradient favoring a unidirectional transfer of drug from the gut to the blood. Blood flow from the GIT drains into the portal vein to the liver before entering the systemic circulation. The first-pass hepatic metabolism has significant implications for oral drug bioavailability. Following penetration through the intestinal membrane, the drug may either enter the systemic blood circulation or the lymphatic system. Generally, lymphatic absorption accounts for only a small proportion of the total drug absorbed, with the exception of drugs with a high oil/water partition coefficient (Mayersohn 1996).

Most drugs are absorbed by a passive diffusion mechanism. The rate of drug diffusion across a membrane is described by Fick's first law of diffusion (Mayersohn 1996):

$$\frac{dQ}{dt} = DAR\left(\frac{C_g - C_b}{X}\right) \tag{9.1}$$

where dQ/dt is the rate of diffusion (or flux); D is the diffusion coefficient of the drug; A is the surface area of the membrane available for drug diffusion; R is the partition coefficient of the drug between the membrane and the aqueous fluid; $C_g$ is the drug concentration at the site of absorption; $C_b$ is the drug concentration in the blood; and X is the thickness of the membrane. The concentration gradient $(C_g - C_b)$ is the driving force for diffusion across the membrane.

Under sink conditions, in which the drug concentration at the site of absorption is much greater than the drug concentration in the blood, the equation may be simplified to a first-order kinetic process (Mayersohn 1996):

$$\frac{dQ}{dt} = KC_g \tag{9.2}$$

Sink conditions apply under physiological conditions due to large blood volume compared with the volume of gut fluid, and rapid clearance of the drug from the site of absorption by systemic circulation. Passive diffusion mechanisms produce a linear relationship between the amount of the drug absorbed and the dose ingested. Active mechanisms produce curvilinear relationships that plateau at high drug doses due to saturation of the membrane (Mayersohn 1996).

Drug absorption by particulate uptake following oral administration may occur through lymphatic tissue (Norris, Puri, and Sinko 1998; Hussain, Jaitley, and Florence 2001). The rate of particle uptake is affected by the physicochemical properties of the drug, including hydrophobicity and particle size (Norris, Puri, and Sinko 1998). A higher rate and extent of particle uptake are observed with reduced particle size (Jani et al. 1990; Jani, McCarthy, and Florence 1992).

The rate of drug absorption depends on the physicochemical properties of the drug, formulation factors, and physiological conditions (Table 9.1). Physicochemical properties of the drug that affect its absorption include the solubility and dissolution rate, oil/water partition coefficient ($K_{o/w}$), degree of ionization (determined by the $pK_a$ of the drug and pH of the biological fluid), and the molecular weight (Mayersohn 1996). Drugs with a higher $K_{o/w}$ have increased membrane permeability. Nonionized drugs also have a

**Table 9.1** Factors Affecting Drug Absorption from Oral Delivery

| Type of Factor | Factor | References |
|---|---|---|
| Drug | Solubility<br>Lipophilicity<br>$pK_a$<br>Oil/water partition coefficient<br>Molecular weight<br>Degree of ionization<br>Crystallinity<br>Polymorphic form | Mayersohn (1996); Norris, Puri, and Sinko (1998); Jani et al. (1990); Jani, McCarthy, and Florence (1992); Hoener and Benet (1996) |
| Dosage form | Type of dosage form<br>Particle size<br>Excipients<br>Compression force of tablets<br>Packing density of capsules | Mayersohn (1996); Nimmerfall and Rosenthaler (1980); Mosharraf and Nyström (1995); Anderberg, Bisrat, and Nyström (1988); Nash (1996) |
| Physiological conditions | pH of gastric fluid<br>Composition of gastric fluid<br>Gastric emptying<br>Intestinal transit<br>Presence of food<br>Disease state<br>Age<br>Diet<br>Body position<br>Emotion state<br>Exercise<br>Pregnancy | Mayersohn (1996); Hoener and Benet (1996) |

greater membrane permeability. Weakly acidic drugs have higher permeability when the pH is less than the $pK_a$, whereas weakly basic drugs have high permeability when the pH is greater than the $pK_a$. However, most drugs are absorbed from the small intestine, regardless of degree of ionization. The dissolution rate from solid dosage forms and the residence times of the drug in different regions of the gastrointestinal tract affect drug absorption (Mayersohn 1996).

Physiological factors influencing drug absorption include pH and composition of gastrointestinal fluids, gastric emptying, intestinal transit, drug degradation and metabolism, presence of food, disease states, and age. The pH of gastrointestinal fluids affects the degree of drug ionization. It varies greatly along the length of the GIT, ranging from 1 to 5 in the stomach, 5.7 to 7.7 in the small intestine, and within 7 to 8 in the large intestine (Mayersohn 1996). The components of GI fluids influence the dissolution rate of drug compounds, such as bile salts (Hoener and Benet 1996).

Gastric emptying has a significant effect on the drug dissolution and absorption of solid dosage forms (Hoener and Benet 1996). Any factor that delays gastric emptying will influence the rate of drug absorption (Mayersohn 1996). The interdigestive migrating motor complex (MMC), commonly known as the "housekeeper wave," occurs approximately every two hours on an empty stomach. The presence of food delays the MMC until the gastric contents are liquid enough to pass the pylorus. Gastric emptying in the presence of food is controlled by a complex variety of mechanical, hormonal, and neural mechanisms. Gastric emptying is influenced by meal volume, presence of acids and nutrients, osmotic pressure, body position (lying on the left or right side), viscosity of liquids, emotional state (depression or stress), gut disease, exercise, and obesity (Mayersohn 1996). Certain drugs, such as metoclopramide, may influence gastric emptying and affect the absorption of other drugs. Large nondisintegrating dosage forms, such as tablets and capsules, rely on the MMC for entry into the small intestine, whereas particles smaller than 5–7 mm may leave the stomach without the MMC (Mayersohn 1996). Thus, there may be a considerable delay in gastric emptying of large dosage forms (Hoener and Benet 1996).

Longer intestinal residence times provide increased opportunity for drug dissolution and absorption (Mayersohn 1996). Intestinal movements may be classified as propulsive and mixing. Propulsive movements produced by peristaltic waves determine the intestinal transit rate (approximately 1–2 cm/sec) and residence time. Intestinal residence times range between one and six hours. Mixing movements are caused by contractions of portions of the small intestine, dividing it into segments. The mixing motion increases drug absorption by increasing the dissolution rate due to agitation and increased contact area between drug and surface epithelium. Factors affecting intestinal transit include the presence of food, presence of other drugs, diet (vegetarian or nonvegetarian), and pregnancy. The influence of food is not as significant on intestinal transit as it is on gastric emptying. The presence of food increases peristaltic activity, motility, and secretion. Food also provides a viscous environment within the intestine, reducing the rate of drug dissolution and diffusion to the absorbing membrane. In addition, drugs may bind to food particles or react with GI fluids secreted in response to food (Mayersohn 1996). Drugs that cause gastrointestinal irritation should be taken with food, whereas acid-labile drugs, or drugs known to exhibit decreased rate and extent of absorption in the presence of food, should be taken on an empty stomach (Robinson, Narducci, and Ueda 1996).

The residence time in the colon is longer and more variable than in other parts of the GIT, varying from several hours to 50–60 hours. The colonic contents are propelled by "mass movement" rather than peristaltic waves. The colon is a potential site of drug absorption due to its long residence time, neutral pH, and low enzymatic activity. However, limitations include small surface area, viscous fluid-like environment, and large colonies of bacteria (Mayersohn 1996).

Drug degradation and metabolism occurring at various sites along the GIT may reduce drug absorption. Enzymatic and acid/base-mediated degradation may occur within the gut fluids, by mucosal cells lining the gut walls, and by microorganisms within the colon (Mayersohn 1996).

The processes involved in drug absorption from solid oral dosage forms include the delivery of the dosage form to the site of absorption, dissolution of the dosage form, drug penetration through membranes of the GIT, and movement away from the site of absorption to the general circulation (Hoener and Benet 1996). The rate of absorption is determined by the slowest step. For poorly water-soluble drugs, dissolution is the rate-limiting step to drug absorption. Dissolution may occur before or after reaching the site of absorption. The rate of dissolution may be described by using the Noyes-Whitney equation:

$$\frac{dc}{dt} = \frac{kDS(C_s - C_t)}{Vh} \tag{9.3}$$

where dc/dt is the rate of dissolution; k is the dissolution constant; D is the coefficient of diffusion; S is the surface area; V is the volume of the dissolution media; h is the thickness of the stagnant diffusion layer; $C_s$ is the drug concentration at the interface; and $C_t$ is the drug concentration in the bulk media.

The effective surface area determines the dissolution rate of drug particles. Reducing the particle size increases the surface area and dissolution rate (Nimmerfall and Rosenthaler 1980; Mosharraf and Nyström 1995; Anderberg, Bisrat, and Nyström 1988). Increased dissolution and bioavailability with particle size reduction have been observed for many poorly soluble drugs, including nitrofurantoin (Conklin and Hailey 1969; Conklin 1978), digoxin (Greenblatt, Smith, and Koch-Weser 1976), phenytoin (Neuvonen, Pentikainen, and Elfving 1977; Neuvonen 1979), griseofulvin (Aoyagi et al. 1982a, 1982b), progesterone, spironolactone, diosmin (Chaumell 1998), danazol (Liversidge and Cundy 1995), naproxen (Liversidge and Conzentino 1995), and benoxaprofen (Smith et al. 1977; Ridolfo et al. 1979). However, aggregation of small particles may increase the effective surface area, resulting in unchanged or reduced dissolution rate and bioavailability (Aguiar, Zelmer, and Kinkel 1967; Jindal et al. 1995). Reduced aggregation of drug particles and increased dissolution rates may be obtained by using solid dispersions (Allen, Kanchick, and Maness 1977; Stavchansky and Gowan 1984) and interactive mixtures (McGinity et al. 1985; Nyström and Westerberg 1986). Drug aggregation and reduced dissolution rate have been observed with increased drug concentration within interactive mixtures (Nilsson, Westerberg, and Nyström 1988; Westerberg and Nyström 1993; Stewart and Alway 1995). The addition of ternary surfactants reduces drug aggregation and increases dissolution rate (Westerberg and Nyström 1993; Stewart and Alway 1995).

The disintegration of the dosage form, presence of other excipients, and manufacturing processes may also affect dissolution rate (Hoener and Benet 1996). The type of lubricant and amount of disintegrant present in tablets affects the disintegration and dissolution rates. Increased compression force may reduce disintegration and dissolution due to the tight binding of the particles within tablets. However, fracture of particles at high compression forces may result in increased disintegration and dissolution rate. High packing density of particles in capsules may reduce the dissolution rate.

Other methods to increase drug absorption involve physicochemical modification to increase the intrinsic solubility of the drug. Such methods include changes in salt formation, degree of solvation, polymorphic form, crystallinity, and complexation (Hoener and Benet 1996).

Oral pharmaceutical preparations are either liquid or solid. Liquid preparations include aqueous solutions, suspensions, and oil-in-water emulsions. Solid preparations include tablets and capsules. Generally, drug bioavailability follows (in descending order): solutions, suspensions, oil-in-water emulsions, capsules, tablets, and modified-release tablets/capsules (Robinson, Narducci, and Ueda 1996). Aqueous solutions are absorbed immediately upon reaching the duodenum, whereas suspensions must dissolve prior to absorption. Tablets and capsules must disintegrate prior to drug dissolution and absorption.

Oral suspensions are useful for delivery of insoluble or poorly soluble drugs and to mask an unpleasant taste from the dissolving drug. Advantages of oral aqueous suspensions are high bioavailability compared with tablets and capsules, ease of swallowing, and greater dose flexibility. However, variations in drug concentration and doses may occur due to nonuniform mixing or difficulty in redispersion (Ofner, Schaare, and Schwartz 1996). The large surface area of suspended particles provides higher availability for absorption. The particle size and size distribution of suspended drugs influences the physical stability (settling rate, resuspendability), product appearance, drug solubility, and bioavailability of suspensions. Suspensions with smaller particle sizes produce more rapid and greater absorption. However, particle size enlargement or crystal growth may occur within suspensions during storage due to Ostwald ripening, temperature fluctuations, and changes in polymorphic form or solvation state. Changes in particle size may be minimized by selection of particles with narrow size ranges, more stable crystalline forms (higher melting points), avoidance of high-energy milling (reduced amorphous content), use of a wetting agent to dissipate the free surface energy of particles, use of protective colloids (gelatin, gums) to form a film barrier around particles to inhibit dissolution and crystal growth, increased viscosity to retard particle dissolution, and avoidance of extreme temperature fluctuations (Nash 1996).

Modified release preparations aim to control the rate-limiting step of drug bioavailability, with the primary objective to maintain plasma drug concentrations in the therapeutic level for prolonged periods (Robinson, Narducci, and Ueda 1996). The advantages of modified release delivery systems include reduced dosage frequency, reduced incidence of undesirable side effects, greater pharmacological activity, and improved patient compliance. Ideally, the rate of drug absorption should equal the rate of drug elimination; however, few delivery systems are able to achieve this goal. Most modified delivery systems provide a first-order drug release rate. Modified delivery systems are generally formulated to provide a rapidly available dose to establish an initial therapeutic plasma level, followed by a controlled-release component to maintain the desired plasma concentration.

Strategies for modified drug release preparations include dissolution-controlled systems, diffusion-controlled systems, ion-exchange resin complexes, and osmotically actuated systems (Robinson, Narducci, and Ueda 1996). The duration of drug release from oral modified release systems is restricted to the limited residence time in the vicinity of the absorption site. Modified release preparations may be monolithic or multiparticulate (pellet) systems. Pelleted dosage forms tend to produce more reproducible transit patterns within the upper GIT than do monolithic dosage forms (Dressman et al. 1994). This reduces variations in drug absorption and minimizes potential side effects and risk of local irritation. Drug release from modified release pellets is influenced by the pellet core formulation, the type and method of coating, and the gastrointestinal conditions into which the drug is released. The particle size and size distribution of pellets greatly affect the drug release profile. Smaller pellets, due to their larger specific surface area, have higher drug release rates (Porter and Ghebre-Sellassie 1994).

## Parenteral Delivery

Parenteral administration is commonly used in reference to injection directly into a body compartment to bypass the protective effects of the skin or mucous membranes, even though the literal definition is any route other than oral administration (Robinson, Narducci, and Ueda 1996). The parenteral route is used for drugs that are poorly absorbed or inactive when administered by other routes. Local or systemic action may be produced, depending on the injection site and formulation. Parenteral delivery is useful in uncooperative, unconscious, or nauseous patients; however, administration by trained personnel is usually required. Injection is generally associated with some degree of pain, and frequent administration is inconvenient.

Parenteral preparations include solutions, suspensions, emulsions, and powders for reconstitution immediately prior to administration. Injectable preparations may be aqueous or nonaqueous and should ideally be sterile, pyrogen-free, isotonic, and nonirritating (Robinson, Narducci, and Ueda 1996). In addition, injectable suspensions should be resuspendable, syringeable, and injectable (Floyd and Jain 1996).

Syringeability and injectability are closely related. Syringeability is the ability of a parenteral solution or suspension to pass easily through a hypodermic needle from the storage vial to injection, such as ease of withdrawal, tendency for clogging and foaming, and accuracy of dose measurements. Injectability is the performance of the suspension during injection, involving the pressure or force required for injection, evenness of flow, aspiration qualities, and tendency of clogging. Both injectability and syringeability are diminished with increased viscosity, density, particle size, and solids concentration. Clogging of the needle may occur due to blockage by a single particle or by the bridging effect of multiple particles (Floyd and Jain 1996). The individual particle size should be no greater than one-third of the needle's internal diameter (Nash 1996).

Intravenous (IV) injection provides an immediate onset of action; other parenteral routes provide slower onset and/or prolonged duration. IV administration yields almost complete drug availability. IV-injected drugs are diluted in the venous system and pass through the heart, and may be eliminated by the lungs prior to entering the general circulation. This is known as the "lung first pass effect." The fraction of drug reaching the desired sites depends on the fraction of arterial blood reaching that site (Oie and Benet 1996). Suspensions should not be administered by IV route (Robinson, Narducci, and Ueda 1996). Nanosuspensions may be injected without the risk of vascular occlusion and pulmonary embolism (Floyd and Jain 1996).

Intramuscular (IM) and subcutaneous (SC) injections are common routes of parenteral administration. IM preparations are injected deep into the skeletal muscles, such as the gluteal (buttocks), deltoid (upper arm), or vastus lateralis (thigh), with injection volumes of 1.5–5 mL. SC preparations are generally small volumes (<2 mL) injected into the loose interstitial tissue beneath the skin of the arm, forearm, thigh, abdomen, or buttocks. Drugs absorbed from IM or SC injection sites enter the venous blood and pass through the heart and lungs prior to entering the general circulation. This results in an initial lag period between the time of injection and time of entry into the general circulation (Oie and Benet 1996). IM and SC administrations are used for drugs that cannot be injected intravenously due to low aqueous solubility and/or when high peak concentrations result in local or systemic side effects (Zuidema et al. 1994).

The rate of drug release and absorption depends on the physicochemical properties of the drug, formulation variables, and injection and physiological factors (Table 9.2) (Feldman 1974; Zuidema, Pieters, and Duchateau 1988). Hydrophilic drugs in aqueous systems are absorbed more rapidly in smaller injection volumes due to greater diffusional potential. Lipophilic drugs in oily vehicles are absorbed more rapidly in smaller injection volumes, again due to greater diffusional potential. Absorption of lipophilic drugs in aqueous vehicles increases with increased injection volume because the drug remains dissolved longer (Zuidema et al. 1994).

The rate-limiting step controlling drug absorption from suspensions is usually the dissolution of a solid drug in biological fluids at the injection site (Feldman 1974). The dissolution rate of suspended particles is governed by the Noyes-Whitney law, equation (9.3),

**Table 9.2** Factors Affecting Drug Release and Absorption from Parenteral Delivery

| Type of Factor | Factor | References |
|---|---|---|
| Drug | Molecular weight<br>$pK_a$<br>Solubility<br>Lipophilicity<br>Oil/water partition coefficient<br>Crystallinity<br>Polymorphic form | Feldman (1974); Zuidema, Pieters, and Duchateau (1988); Zuidema et al. (1994); Floyd and Jain (1996) |
| Dosage form | Drug concentration<br>Particle size<br>Type of vehicle (aqueous/nonaqueous)<br>Composition of vehicle (cosolvents, surfactants)<br>pH of vehicle<br>Viscosity<br>Excipients<br>Osmolality | Feldman (1974); Zuidema, Pieters, and Duchateau (1988); Zuidema et al. (1994); Nash (1996); Floyd and Jain (1996) |
| Administration | Injection site<br>Injection depth<br>Injection volume<br>Injection technique<br>Injection depot shape<br>Needle gauge<br>Needle length | Zuidema, Pieters, and Duchateau (1988); Zuidema et al. (1994); Oie and Benet (1996); Robinson, Narducci, and Ueda (1996); Floyd and Jain (1996) |
| Physiological conditions | Blood supply<br>Body movement<br>Tissue transport<br>Lymphatic transport | Zuidema, Pieters, and Duchateau (1988); Zuidema et al. 1994) |

which states that the dissolution rate is proportional to the surface area, diffusion coefficient, and concentration gradient (Zuidema, Pieters, and Duchateau 1988). Suspensions that contain smaller particles have faster dissolution rates, due to the increased surface area of the particles (Hirano and Yamada 1982). Smaller particles produced higher bioavailability for some drugs, such as medroxyprogesterone (Antal et al. 1989). However, particle aggregation may reduce the dissolution rate, due to increased effective surface area, increased viscosity, and altered rheology. Suspensions of larger particles have slower dissolution rates and provide sustained drug release and prolonged action (Chien 1992c; Butterstein and Castracane 2000). Increasing drug concentration may lead to aggregation and reduced dissolution and absorption rates. Excipients, such as surfactants, may reduce aggregation and increase the dissolution and absorption rates. Altering the suspension vehicle may affect the dissolution and absorption rates. Increased viscosity reduces the diffusion coefficient and dissolution rate. It is important for the drug to remain in solution at the injection site for absorption. This is not a problem for oily systems because the rate of clearance of the oily vehicle is generally slower than the rate of absorption of the drug. Precipitation of drugs may occur at the injection site following the absorption of the aqueous vehicle and change in pH (Zuidema et al. 1994).

Absorption generally occurs by passive diffusion. Other mechanisms for drug absorption from parenteral administration include phagocytosis by macrophages or lymphatic transportation. Drug uptake by the lymphatic system is mostly achieved by SC or intraperitoneal (IP) injection. The absorption pathway of injected drugs depends on injection depth, lipophilicity, and size of particles or carrier. Molecules smaller than 5 kDa favor absorption into capillaries. Molecules larger than 16 kDa and particles or liposomes larger than 200 nm are preferentially drained into the lymphatic system (Zuidema et al. 1994).

The intensity and duration of drug activity depend upon the physicochemical and pharmacokinetic properties of the drug, degree of plasma protein binding, extent of distribution throughout the body, and rate of elimination by metabolism and/or excretion (Robinson, Narducci, and Ueda 1996). Depot injections are controlled release formulations (either aqueous suspensions or oleaginous solutions) injected into subcutaneous or muscular tissues. The depot formed at the injection site acts as a drug reservoir providing prolonged drug release and duration of action (Chien 1992c). Depot injections provide constant and sustained drug levels; reduce the frequency of injection, dose required and side effects; and improve patient compliance. The processes involved in the drug release from an IM drug depot include diffusion of water to the depot, dissolution of suspended drug particles, diffusion in the dispersing agent, drug transfer from oily vehicles into water phase, and diffusion away from the depot to the systemic circulation (Zuidema 1988). The rate of drug release from depot formulations is controlled by dissolution, adsorption, encapsulation, and esterification. Reduced dissolution may be provided by increasing the size of suspended particles (Chien 1992c). Other types of parenteral preparations include emulsions and liposomes (Robinson, Narducci, and Ueda 1996; Floyd and Jain 1996; Zuidema et al. 1994).

## Respiratory Delivery

The respiratory route of administration by inhalation is a well-established method of administration for various drugs (Clarke and Newman 1984). Both local and systemic therapeutic action may be provided by respiratory delivery. Although inhalation is used mainly for respiratory diseases, investigations are under way for pulmonary delivery of peptides and proteins (Smith 1997; Patton, Bukar, and Nagarajan 1999) and controlled drug release (Byron 1986; Gonzalez-Rothi and Schrier 1995; Zeng, Martin, and Marriot 1995). The respiratory route provides advantages over oral and parenteral routes. Smaller doses may be used to provide a local effect, thereby reducing the risk of systemic side effects or toxicity. Compared with oral administration, pulmonary delivery provides more rapid onset of action and avoids gastrointestinal degradation (enzymatic or chemical) and hepatic first-pass elimination. Inhalation is non-invasive and nontraumatic, and it avoids the risk of transmitting blood-borne pathogens, encountered with parenteral injections.

Effective drug delivery requires deposition into the pulmonary regions of the lungs, enabling absorption prior to clearance. The respiratory tract is a series of airways, subdividing from the trachea to bronchi to bronchioles and terminating at the alveolar sacs. From the trachea to the alveoli, the airway caliber diameter decreases from about 1.8 cm to 0.04 cm; but the surface area increases to about 140 $m^2$ (Hickey and Thompson 1992). The primary function of the respiratory tract is to facilitate gas exchange. The large surface area and two-cell thickness of the alveolar region as well as the high blood flow provide for a useful route of drug administration.

Particle size and distribution are key elements in aerosol drug delivery and efficacy of drugs delivered by the pulmonary route. Deposition in the respiratory tract occurs by five main physical mechanisms: inertial impaction, sedimentation, diffusion, interception, and electrostatic deposition (Gonda 1990). Deposition by inertial impaction occurs due to the momentum of the aerosol particles. Particles with higher mass or velocity have longer stopping distances and increased chances of deposition on the walls of the respiratory tract.

Generally, larger particles (above 5 μm) with high velocity are deposited in the back of the mouth and upper airways by inertial deposition (Gonda 1992). Deposition by sedimentation is governed by Stokes' law. Smaller particles (0.5 to 3 μm) are deposited in the bronchial and alveolar regions by sedimentation (Brain and Blanchard 1993). Increased sedimentation occurs during either breathholding or slow tidal breathing (Gonda 1992). Deposition by diffusion is due to Brownian motion, caused by constant random collisions of gas particles with small aerosol particles. Aerosol deposition by diffusion is independent of particle density but increases with decreasing particle size. It is dependent upon residence time and enhanced by breathholding. In general, inertial impaction and sedimentation dominate the deposition of particles larger than 1 μm and diffusion dominates the deposition of particles smaller than 0.1 μm. For the size range between 0.1 and 1 μm, both sedimentation and diffusion are important (Brain and Blanchard 1993). Deposition by interception generally occurs for elongated particles. Electrostatic deposition occurs for charged particles.

Drug deposition is affected by the aerosol characteristics, breathing patterns, and airway caliber (Newman and Clarke 1983). The most important characteristic of the aerosol is generally considered to be given by the aerodynamic diameter, $d_{ae}$, defined as the equivalent diameter of a sphere of unit density that has the same settling velocity in still air as the particle in question (Gonda 1992):

$$d_{ae} = d\sqrt{\frac{\rho}{\rho_0}} \tag{9.4}$$

where d is the diameter of the sphere; ρ is the density of the sphere; and $\rho_0$ is unit density [see Chapter 6, eq (6.4)]. Aerosol droplet or particle size is generally polydisperse, following a lognormal distribution, which enables characterization by the mass median aerodynamic diameter (MMAD) and the geometric standard deviation (GSD). Greater aerosol deposition into the lung periphery occurs with lower MMAD and higher proportion of particles below 5 μm (Dolovich 1992). Size enlargement of aerosol droplets or particles may occur in the respiratory tract, due to hygroscopic growth (Gonda 1990; Gonda and Byron 1978; Hickey and Martonen 1993). Other aerosol characteristics, such as particle shape, influence aerosol deposition.

The breathing pattern of the patient may influence aerosol deposition (Newman and Clarke 1983; Martonen and Katz 1993). Factors include inspiratory airflow rate, respiratory volume, respiratory frequency, and breathholding duration at the end of inspiration (Gonda and Byron 1978; Martonen and Katz 1993; Newman 1985). Airway geometry affects aerosol deposition (Newman 1985). Variations in airway geometry are due to differences in gender, lung volume, age, and disease state (Brain and Blanchard 1993; Newhouse and Dolovich 1986).

Drug absorption in the respiratory tract occurs mainly by passive diffusion (Gonda 1990), by which absorption is determined by molecular size and lipophilicity. Other absorption routes include transport through aqueous pores, carrier-mediated active transport processes, and lymphatic transport (Gonda 1990; Thompson 1992).

Clearance of foreign matter from the conducting airways occurs primarily by the mucociliary escalator, and clearance from the alveolar regions occurs predominantly by alveolar macrophage phagocytosis or enzymatic metabolism. Phagocytosis is followed by subsequent removal of foreign matter by either the mucocilary escalator or the lymphatic system (Niven 1992). Mucociliary clearance from the lower respiratory tract is completed within 24 hours, but clearance by macrophage phagocytosis is slower (Morén 1993). Pulmonary metabolism is generally considered lower than hepatic metabolism. However, the lung serves as an important organ of metabolism for endogenous substances in the systemic blood system (Thompson 1992).

Various aerosol devices are used for respiratory drug administration (Crowder et al. 2001; Dunbar, Hickey, and Holzner 1998a). Nebulizers produce liquid aerosol droplets by Bernoulli's effect or high-frequency vibrations from a piezoelectric crystal (McCallion et al. 1996; Taylor and McCallion 1997; Flament, Leterme, and Gayot 1995). Handheld aqueous systems are more convenient and portable than conventional nebulizers (Newman et al. 1996; Hickey and Dunbar 1997; Schuster et al. 1997). Metered dose inhalers (MDI) deliver liquid droplets containing drug using liquefied gas propellents (Hickey and Dunbar 1997). Dry powder inhalers (DPI) deliver drug particles using the patient's inspiratory airflow for powder dispersion and deaggregation (Dunbar, Hickey, and Holzner 1998a; Ganderton 1992; Prime et al, 1997).

Aerosol dispersion and drug absorption is affected by physicochemical properties of the drug, delivery device and formulation factors, and physiological factors (Table 9.3). DPI powder formulations require drug particles to have aerodynamic diameters below 5 µm for lung deposition. Respirable-sized drug particles may be prepared by either size reduction through milling or particle construction through condensation, evaporation, or precipitation (Hickey et al. 1994; Sacchetti and van Oort 1996). Such particles generally exhibit poor flow properties due to their high interparticle forces. Formulation strategies to improve the flowability of respirable particles include the controlled agglomeration of drug particles or adhesion onto excipient carrier particles in the form of interactive mixtures (Hersey 1975). Aerosol dispersion of the drug particles from aggregates or interactive mixtures is required for lung deposition (Ganderton 1992). The aerosol dispersion depends upon the particle interactions within the powder formulation and the mechanical forces of dispersion from the device. Strong interparticulate forces within the powder formulation may lead to poor efficiency (Byron 1986). Different mechanisms of aerosol dispersion are provided by the various DPI devices (Crowder et al. 2001; Dunbar, Hickey, and Holzner 1998a). Factors affecting particle adhesion (see Chapter 5) and aerosol dispersion of DPI formulations include physicochemical properties of the drug and carrier, such as size, shape, surface roughness, chemical composition, polymorphic form and crystalline state, the drug-carrier ratio, and the presence of ternary components (Crowder et al. 2001; Dunbar, Hickey, and Holzner 1998a).

Drug particle size affects aerosol dispersion: Smaller particles produce lower fine-particle fraction (FPF) of drug-alone formula-

**Table 9.3** Factors Affecting Drug Deposition and Absorption from Dry Powder Inhalers

| Type of Factor | Factor | References |
|---|---|---|
| Drug | Chemical composition<br>Molecular weight<br>Solubility<br>Lipophilicity<br>$pK_a$<br>Oil/water partition coefficient<br>Crystallinity<br>Polymorphic form<br>Electrostatic properties | Niven (1992); Gonda (1990); Thompson (1992) |
| Formulation | Type of formulation (drug aggregates or interactive mixtures)<br>Particle size<br>Particle shape<br>Particle density<br>Carrier size<br>Drug concentration<br>Ternary components | Newman and Clarke (1983); Byron (1986); Dunbar, Hickey, and Holzner (1998a); Crowder et al. (2001) |
| Device | Type of device (active/passive)<br>Inhaler design<br>Energy input<br>Specific resistance | Dunbar, Hickey, and Holzner (1998a); Crowder et al. (2001) |
| Physiological conditions | Airflow rate<br>Airway geometry<br>Mucociliary clearance | Newman and Clarke (1983); Byron (1986); Martonen and Katz (1993); Brain and Blanchard (1993); Thompson (1992); Morén (1993) |

tions, due to higher cohesion, whereas larger particles are less cohesive but more likely to deposit in the upper stage due to inertial impaction (Chew and Chan 1999; Chew, Bagster, and Chan 2000). Surface modification by adhesion of nanoparticles onto the drug particles may increase aerosol dispersion of drug-alone and carrier-based formulations (Kawashima et al. 1998a, 1998b). The particle size of carrier particles can influence aerosol dispersion. Increased drug deposition is generally observed with smaller carrier size (Kassem, Ho, and Ganderton 1989; Braun, Oschmann, and Schmidt 1996; Steckel and Müller 1997) and increased proportion of fine particles (Mackin, Rowley, and Fletcher 1997; Srichana, Martin, and Marriott 1998). Increased drug concentration may reduce FPF due to drug aggregation (Braun, Oschmann, and Schmidt 1996; Steckel and Müller 1997). The addition of fine ternary components has been shown to increase the FPF of various drugs (Ganderton 1992; Staniforth 1996; Lucas, Anderson, and Staniforth 1998; Zeng et al. 1998). Possible explanations for the mechanism of action of ternary components include the saturation of active sites on the carrier, electrostatic interactions, and drug redistribution on the ternary component. Other factors affecting adhesion include environmental conditions; the duration of particulate contact; and initial contact velocity, which influences the adhesion of particles within the powder mix and to surrounding surfaces, such as the device walls (Hickey et al. 1994).

## Nasal Delivery

The nasal route of drug administration may be used for topical and systemic action (Kublik and Vidgren 1998). Advantages of nasal delivery include rapid absorption, high bioavailability, fast onset, avoidance of gastrointestinal metabolism and hepatic first-pass metabolism, noninvasiveness, improved patient compliance, and reduced risk of infection. Limitations of nasal delivery include possible poor absorption and metabolic degradation of some drugs in the nasal cavity. Nasal delivery may not be suitable for drugs with poor aqueous solubility, drugs that require sustaining blood levels, or for chronic conditions (Behl et al. 1998).

The nasal passage extends from the nostrils to the nasopharynx. The main nasal passages (superior, middle, and inferior turbinates) are highly vascularized and ciliated. The turbinates consist of projections that extend into the nasal cavity and increase the surface area. Higher airflow resistance and increased turbulence is observed in the turbinates (Kublik and Vidgren 1998). The nasal cavity volume is about 20 mL, and its total surface area is around 180

$cm^2$. The thickness of the mucosa is 2–4 mm (Behl et al. 1998). The functions of the nose include olfaction, chemical sensation, defense by using various immunological secretions and mucocilary clearance, filtration, and warmth and humidification of inhaled air (Jones 2001). Mucociliary clearance limits the residence time of formulations to remain in the nasal cavity (Jones 2001). Drug absorption occurs from the mucosal surfaces in the posterior region of the nasal cavity (Robinson, Narducci, and Ueda 1996). The permeability of the nasal mucosa exceeds the permeability of buccal, duodenal, jejunal, ileal, colonic, and rectal tissues (Donovan and Huang 1998). The nasal epithelium is termed to be a leaky mucosal tissue. Nasal drug absorption may occur by transcellular and paracellular passive transport, carrier-mediated transport, and transcytosis (Kublik and Vidgren 1998).

Drug bioavailability following nasal administration is affected by physicochemical properties of the drug, formulation factors, administration technique, and physiological conditions (Table 9.4). The deposition site influences drug absorption and bioavailability. Nasal deposition mainly depends on particle or droplet size, airflow rate, and nasal geometry. Inhaled particles are deposited mainly by inertial impaction. Translocation by mucociliary clearance results in a secondary deposition of the drug (Kublik and Vidgren 1998). The ideal volume for nasal administration is 25–200 μL per nostril. Larger volumes will drain out of the nose (Behl et al. 1998).

Nasal absorption and bioavailability are affected by the physicochemical properties of the drug, such as molecular weight (MW), solubility and dissolution rate, $pK_a$ and partition coefficient, degree of ionization, particle size and morphology, and polymorphic state (Behl et al. 1998). The nasal absorption of drugs with MW less than 300 Da is not significantly influenced by the other physicochemical properties of the drug. Increased molecular weight above 300 Da reduces absorption (Hussain 1998). Nasal administration of aqueous solutions of propranolol have produced identical blood levels to those obtained by IV injection (Chien and Chang 1985). Low bioavailability is observed for drugs larger than 1000 Da. Penetration enhancers may be used to increase absorption and bioavailability. Particulate uptake has been observed from the nasal mucosa by nasal-associated lymphoid tissues (NALT) (Donovan and Huang 1998).

**Table 9.4** Factors Affecting Drug Absorption from Nasal Administration

| Type of Factor | Factor | References |
|---|---|---|
| Drug | Molecular weight<br>Solubility<br>Lipophilicity<br>$pK_a$<br>Oil/water partition coefficient<br>Crystallinity<br>Polymorphic form | Behl et al. (1998) |
| Dosage form and formulation | Dose<br>Drug concentration<br>Particle size<br>Particle shape<br>Particle density<br>Viscosity<br>pH<br>Tonicity<br>Osmolality<br>Excipients | Robinson, Narducci, and Ueda (1996); Behl et al. (1998); Kublik and Vidgren (1998) |
| Administration | Type of device<br>Position of head<br>Droplet size<br>Volume<br>Deposition site | Robinson, Narducci, and Ueda (1996); Kublik and Vidgren (1998) |
| Physiological conditions | Airflow rate<br>Nasal geometry<br>Mucociliary clearance<br>Disease state<br>Emotional state<br>Infection<br>Blood flow | Robinson, Narducci, and Ueda (1996); Behl et al. (1998); Kublik and Vidgren (1998) |

The rate of absorption also depends on physiological factors, such as rate of nasal secretion, ciliary movement, and metabolism. Reduced bioavailability is observed with increased rate of nasal secretion and mucociliary clearance (Hussain 1998).

Drugs are eliminated from the nasal cavity via the mucociliary clearance. Nasal clearance occurs in three phases (Kublik and

Vidgren 1998): The initial phase involves a very rapid clearance, due to swallowing or run-through of the formulation into the pharynx. It occurs within the first minute of administration and is mainly applicable for liquid formulations and large volume applications. The second phase involves the progressive clearance of material deposited in the ciliated area of the main nasal passages. It occurs within 15 to 30 minutes after administration. The third clearance phase is the slow clearance from nonciliated areas of the nose; this phase lasts several hours. The overall clearance of liquid formulations is approximately 30 minutes with healthy subjects. The contact time may be prolonged by use of bioadhesive excipients or thickening agents, such as methylcellulose (Kublik and Vidgren 1998).

Drug metabolism in the nasal cavity is minimal, due to the fast absorption rate and relatively short exposure time of enzymes and the low levels of enzymes that are present (Hussain 1998). The effect on absorption for most compounds, except peptides, is insignificant.

The delivery systems used for nasal delivery include nose drops, aqueous sprays, MDIs, powders, and gels (Kublik and Vidgren 1998; Behl et al. 1998). Liquid aqueous formulations are the most commonly used dosage form for nasal administration. Major drawbacks include microbiological stability, reduced chemical stability, and short residence times in the nasal cavity. Nasal powders and gels are less commonly used, although both may prolong contact time with the nasal mucosa. Nasal suspensions are not commonly used due to the limited amount of water available in the nasal cavity for dissolution.

For particulate nasal products (dry powders or suspensions), the dissolution rate of the drug affects the absorption rate. Particles deposited in the nasal cavity generally must be dissolved prior to absorption. If drug particles are cleared before absorption, bioavailability is reduced. The particle size, morphology, and polymorphic state affect dissolution and absorption (Behl et al. 1998).

## Transdermal and Topical Delivery

Drugs delivered by the topical route produce local pharmacological action within skin tissues. The transdermal route delivers drugs through the skin to produce a regional or systemic pharmacological action at a location away from the site of application (Ramachandran and Fleisher 2000). Transdermal delivery avoids chemical degradation and drug metabolism in the gastrointestinal tract as well as the hepatic first-pass effect. It eliminates the risks and inconveniences of parenteral administration and improves patient compliance. Transdermal delivery is useful for drugs with narrow therapeutic indices or short biological half-lives. Drugs delivered transdermally must penetrate the skin in sufficient quantities to exert a systemic effect without being affected by enzymes in the epidermis (Robinson, Narducci, and Ueda 1996). Therapeutic potency is required with doses of less than 10 mg.

The skin acts as a protective barrier with sensory and immunological functions (Foldvari 2000). Its primary functions include containment of tissues and organs, protection from dangers (microbacterial, chemical, radiation, thermal, and electrical), environmental sensing (tactile/pressure, pain, and thermal), heat regulation, metabolism, disposal of biochemical waste through secretions, and blood pressure regulation (Flynn 1996). The surface area of the skin is approximately 2 $m^2$ (Hadgraft 2001). The pH of the skin ranges between 4.8 and 6.0 (Ramachandran and Fleisher 2000).

The skin consists of three main layers—the stratum corneum, epidermis, and dermis. The stratum corneum is the outermost layer of the skin and provides an effective barrier to penetration (Hadgraft 2001). The stratum corneum consists of layers of nonviable keratinocytes with intercellular lipid phase containing ceramides, free sterols, free fatty acids, triglycerides, sterol esters, and cholesterol sulphate, arranged in a bilayer format (Ramachandran and Fleisher 2000; Hadgraft 2001). The stratum corneum is approximately 10–25 μm thick (Burkoth et al. 1999). The keratinocytes are constantly shed and replaced by differentiating kerotinocytes migrating from the viable epidermis (Foldvari 2000). The stratum corneum and viable epidermis possess minimal, if any, vascularization. The dermis is a highly vascularized layer of structural collagen matrix, elastin, and ground substance. It is embedded with appendages, such as hair follicles, sebaceous glands, and sweat glands. It also contains lymphatic vessels and nerve endings. Fibroblasts, macrophages, mast cells, leukocytes, and other cells are scattered thoughout the dermis (Foldvari 2000). The bloodflow rate is about 0.05 mL/min/$mm^3$ and is altered by physiological and psychological responses, such as exercise, temperature, and shock (Burkoth et al. 1999). The metabolism of substrates and foreign substances occurs in the dermis.

The stratum corneum provides the rate-limiting barrier to drug absorption through the skin (Robinson, Narducci, and Ueda 1996; Foldvari 2000). Drug diffusion through the viable epidermis and dermis is faster than through the stratum corneum. The viable tissue and capillary walls are more permeable than the stratum corneum (Ramachandran and Fleisher 2000). Drug absorption through the stratum corneum occurs by intercellular, transcellular, and appendageal transport. The appendageal route does not significantly contribute to drug penetration, due to the low surface area occupied by the appendages (0.1 percent of the total skin surface). However, appendageal transport may affect the onset of the absorption process (Ramachandran and Fleisher 2000; Foldvari 2000; Hadgraft 2001). The intercellular route is generally assumed to be the major route for drug transport, occurring by repeated partition and diffusion across the bilayers of the stratum corneum (Hadgraft 2001). Percutaneous drug absorption is slower and more selective than gastrointestinal absorption (Robinson, Narducci, and Ueda 1996). Passive diffusion is governed by Fick's first law of diffusion, which states that the concentration gradient is the driving force for diffusion (Foldvari 2000).

The rate and extent of absorption is determined by the physicochemical properties of the drug, the dosage form, and skin condition (Table 9.5). The physicochemical properties of a drug that affect transdermal absorption include potency, solubility, lipophilicity, polarity, octanol-water coefficient (log P), $pK_a$, degree of ionization, crystallinity, and molecular weight (Ramachandran and Fleisher 2000; Hadgraft 2001). Highly lipophilic drugs penetrate the skin easily (Oie and Benet 1996). Generally, compounds with a molecular weight greater than 500 Da cannot penetrate the skin (Foldvari 2000). Transdermal drug absorption may be increased by various methods, such as the use of hydration, penetration enhancers, prodrugs, phonophoresis, or iontophoresis (Ramachandran and Fleisher 2000; Foldvari 2000; Hadgraft 2001; Burkoth et al. 1999).

Formulation factors affecting transdermal drug absorption include the type of dosage form, the nature of the vehicle, the presence of surfactants, and other excipients (Foldvari 2000). Physiologi-

**Table 9.5** Factors Affecting Drug Absorption from Transdermal Delivery

| Type of Factor | Factor | References |
|---|---|---|
| Drug | Potency<br>Molecular weight<br>Solubility<br>Lipophilicity<br>Polarity<br>$pK_a$<br>Log P<br>Degree of ionization<br>Crystallinity<br>Polymorphic form | Ramachandran and Fleisher (2000); Hadgraft (2001); Oie and Benet (1996); Foldvari (2000) |
| Formulation | Type of vehicle<br>Excipients<br>Surfactants<br>Penetration enhancers<br>Particle size<br>Particle density<br>Particle strength | Foldvari (2000); Robinson, Narducci, and Ueda (1996); Burkoth et al. (1999) |
| Physiological conditions | Anatomical location<br>Degree of hydration<br>Age<br>Surface temperature<br>Surface sebum film | Ramachandran and Fleisher (2000) |

cal factors affecting transdermal absorption include age of skin, anatomical location, degree of skin hydration, changes in microcirculation, surface temperature, and surface sebum film (Ramachandran and Fleisher 2000).

Transdermal liquid preparations include solutions, suspensions, and emulsions. Semisolid preparations include ointments, creams, gels, and pastes. Skin permeability follows (in descending order): oily base, water-in-oil emulsions, and oil-in-water emulsions (Robinson, Narducci and Ueda 1996). Other transdermal dosage forms include transdermal patches (Robinson, Narducci, and Ueda 1996; Ramachandran and Fleisher 2000), topical aerosols (Morgan et al. 1998), and intradermal powder (needle-free) injections (Burkoth et al. 1999).

Intradermal powder injections use a supersonic gas flow to accelerate drug particles through the skin at high velocities (300–900 m/s). The entrained drug particles penetrate through the stratum corneum and deposit in the epidermis. The depth of penetration is directly related to the particle size, density, and velocity, which is controlled by nozzle design and gas pressure. The injected drug particles dissolve within the epidermis and diffuse to their intended site of action, either local or systemic. Intradermal powder injection was shown to give results similar to subcutaneous needle injection (Burkoth et al. 1999). The needle-free injection is painless and allows patient self-administration. Additionally, powder formulations provide greater storage stability. Local tissue reactions include mild and transient erythema. Physicochemical properties of the drug and formulation may affect drug penetration and absorption. The particle size and size distribution affects drug penetration. Drug penetration occurs with particles larger than 20 μm. Particles with reduced density (porous or hollow particles) have reduced penetration. Particle shape and surface morphology do not appear to affect drug delivery. The particles must be strong enough to withstand particle–particle and particle–wall collisions in the helium gas jet. Physiological factors affecting drug absorption include dissolution characteristics, pH profile during dissolution, local osmotic pressure, local tissue binding, and drug metabolism (Burkoth et al. 1999).

## Ocular Delivery

Topical administration to the eyes is a common route of drug delivery for ocular diseases and diagnostics. The sites of action for most ophthalmic drugs are located within the eye. The cornea is the main barrier for ocular drug absorption. Most ocular preparations exhibit low bioavailability, typically less than 5 percent (Jarvinen, Jarvinen, and Urtti 1995). The major disadvantage of ophthalmic delivery is the relatively frequent dosing regimen, resulting from rapid clearance, low residence time, and low bioavailability.

Common ophthalmic preparations include eye drops or ointments. Eye drops are usually aqueous, sterile, buffered solutions that contain preservatives and viscosity-enhancing agents. Suspensions may be used for insoluble, unstable drugs or when prolonged action is required. The rate and extent of drug absorption is determined by the particle size. Eye ointments contain drug dissolved or suspended in a petroleum-based vehicle. Ointments are used to prolong drug release and contact time, but may result in blurred vision (Robinson, Narducci, and Ueda 1996).

Drug clearance from the lacrimal fluid occurs by solution drainage, lacrimation and nonproductive absorption to the conjunctiva (Jarvinen, Jarvinen, and Urtti 1995). The normal holding capacity of the lower cul-de-sac for tears is 7 μL. Approximately 30 μL may be held momentarily without blinking. The majority of a drop (around 50 μL) is lost due to spillage and drainage into the nasolacrimal duct, which eventually leads to the gastrointestinal tract (Bapatia and Hecht 1996). Tears result in a rapid dilution of drug and limited contact time for drug absorption.

The cornea is an optically transparent tissue that acts as the principal refractive element of the eye. The corneal thickness is 0.5–0.7 mm, and the surface area of the cornea is 1 $cm^2$ (Jarvinen, Jarvinen, and Urtti 1995). The cornea consists of three layers—a thick aqueous stroma sandwiched between the lipid epithelium and endothelium layers. A drug must have both lipophilic and hydrophilic properties to penetrate the cornea (Bapatia and Hecht 1996). Drug transport across the cornea occurs via transcellular and paracellular pathways, predominantly by passive diffusion. The corneal epithelial is less permeable than the intestinal, nasal, or pulmonary epithelial tissue; but more permeable than the stratum corneum. Absorbed drugs may be eliminated by the aqueous humor turnover, blood circulation, or enzymatic metabolism (Jarvinen, Jarvinen, and Urtti 1995).

Absorbed drugs generally penetrate the anterior segments of the eye (cornea, anterior chamber, iris, crystalline lens, and ciliary body). Penetration to the posterior segments of the eye (vitreous body, retina, and choroids) is poor. Thus, drug administration to the posterior eye involves direct injection into the vitreous cavity or systemic administration. New approaches for drug delivery to the posterior eye include microparticle carriers (liposomes, microspheres, and nanospheres), polymeric vitreal implants, transscleral systems, and targeted delivery via systemic circulation (Ogura 2001).

The rate and extent of corneal absorption is determined by the physicochemical properties of the drug and formulation variables. In general, ocular drug absorption is affected by the physicochemical properties of the drug, such as lipophilicity, octanol/water partition coefficient (log P), solubility, molecular size, and degree of ionization (Jarvinen, Jarvinen, and Urtti 1995; Bapatia and Hecht 1996). Reduced particle size has been shown to increase ophthalmic bioavailability of dexamethasone suspensions due to increased dissolution (Schoenwald and Stewart 1980). However, a change in absorption was not observed with prednisolone suspension, possibly due to permeability rate–limited absorption (Bisrat et al. 1992). Reduced particle size was shown to increase ophthalmic bioavailability of disulfiram solid dispersions (Nabekura et al. 2000).

Drug absorption may be affected by the physiological nature of the eye. Factors include the limited capacity of holding dosage forms, tear fluid and aqueous humor secretion and drainage rates, residence time and spillage, blinking rate, and reflex tearing (Bapatia and Hecht 1996). Increased ocular absorption has been investigated using aqueous suspensions; penetration enhancers; specialized devices (collagen shields, iontophoresis, and pumps); ion-exchange resins; microspheres; inserts; prodrugs; and mucoadhesives (Bapatia and Hecht 1996; Saettone and Salminen 1995).

## Otic Delivery

Drug delivery by the otic route provides local action to the external ear exclusively. Commonly used drugs administered by the otic route include antibiotics, antiinflammatories, and anesthetics. Systemic drug administration is necessary for conditions affecting the middle and inner ear (Robinson, Narducci, and Ueda 1996).

Factors affecting drug absorption from the ear are identical to those governing transdermal drug absorption. However, drug absorption is similar to that observed for mucous membranes when the tympanic membrane (ear drum) is perforated (Robinson, Narducci, and Ueda 1996).

Otic preparations are generally nonaqueous, water-miscible drops containing glycerol, and propylene glycol vehicles. Surfactants are added to promote mixing with oily secretions of the sebaceous and cerumen glands. Suspensions less desirable as insoluble particles may compact with the ear cerumen (Robinson, Narducci, and Ueda 1996).

## Buccal and Sublingual Delivery

The sublingual and buccal surfaces of the oral mucosa have a high degree of vascularization with relatively little keratinization. This provides great potential for drug delivery. The buccal mucosa has a thickness of 600 μm, a surface area of 50 $cm^2$, and a turnover of 13 days. The sublingual mucosa has a thickness of 200 μm, a surface area of 26 $cm^2$, and a turnover of 20 days. The sublingual membrane is more permeable than the buccal membrane (Burkoth et al. 1999).

Sublingual drug absorption is rapid, whereas buccal absorption allows prolonged retention of the dosage form. Advantages of oral mucosal drug delivery include avoidance of enzymatic degradation in the gastric fluids and avoidance of hepatic first-pass metabolism. Limitations include the difficulty of maintaining dosage forms at the sublingual or buccal sites, lack of penetration of some drugs, and the small surface area available for drug absorption. Preparations used for buccal or sublingual delivery include tablets, lozenges, liquid capsules, and aerosols (Lamey and Lewis 1990).

The oral mucosal membrane forms an effective barrier by the presence of intact stratified epithelium, similar to skin. The oral mucosa is lubricated and protected with saliva (pH of 6.2–7.4). The saliva increases the permeability of the oral mucosa by surface hydration. However, limited fluid volume is available for drug dissolution. The dissolution rate of drugs is affected by the salivary flow rate (Lamey and Lewis 1990).

Drug transport across the oral mucosa occurs through the transcellular and paracellular routes. Absorption occurs mainly by passive diffusion. Drug absorption is affected by the lipid solubility, partition coefficient, molecular weight, solubility at site of absorption, and degree of ionization (Chien 1992b; Lamey and Lewis 1990). High potency is required for effective drug administration from buccal and sublingual routes, due to the limited surface area for absorption (Lamey and Lewis 1990).

## Rectal Delivery

The rectal route of drug administration may be used for local and systemic action. It is used as an alternative to oral administration in patients with nausea or vomiting (Robinson, Narducci, and Ueda 1996). Advantages of rectal delivery include rapid absorption, avoidance of first-pass metabolism (if the drug is delivered to the lower rectum), and avoidance of gastrointestinal degradation and irritation. Disadvantages include the limited surface area for drug absorption and limited fluid for drug dissolution (Mackay, Phillips, and Hastewell 1997).

The rectum is the terminal segment of the large intestine. It is 15–20 cm in length and contains approximately 2–3 mL of mucous fluid secretions. The rectum fluid has a pH of 7–8, with no buffering capacity or enzymatic activity. The surface area available for absorption is 200–400 $cm^2$. The site of absorption within the rectum (depth of rectal insertion) determines the degree of first-pass metabolism. The superior hemorrhoidal vein drains directly into the portal vein, whereas the middle and inferior hemorrhoidal veins drain into the systemic circulation (Robinson, Narducci, and Ueda 1996). Suppositories tend to move upward toward the superior hemorrhoidal vein (Oie and Benet 1996).

The physical barriers to rectal drug absorption include the mucus, unstirred layer, and mucosal wall (Edwards 1997). Drug absorption across the rectal mucosa generally occurs by passive diffusion. In addition, enzymatic metabolism by the bacterial flora may affect drug absorption. Lipid soluble drugs are rapidly absorbed, whereas ionic and hydrophilic drugs are poorly absorbed (Chien 1992a). Physicochemical properties of the drug that affect absorption include solubility, $pK_a$, degree of ionization, lipophilicity, molecular size, and polarity (Edwards 1997). Physiological factors affecting drug absorption include rectal motility (residence time), gender, age, blood flow rate, diet, disease state, diurnal variations, luminal pressure, presence of fecal material, and other drug therapy (Robinson, Narducci, and Ueda 1996; Edwards 1997).

Preparations used for rectal administration include suppositories and enemas. Enemas are solutions or suspensions used for local and systemic action (Robinson, Narducci, and Ueda 1996). Suppositories dissolve in the rectal secretions and the drug is absorbed in the rectal mucosa. Generally, lipophilic drugs are formulated in hydrophilic bases, and water-soluble drugs are formulated in lipophilic bases. Suppositories should be retained in the rectum for at least 20 to 30 minutes to ensure complete melting and dissolution of the base and drug release. The particle size of drug dispersed in suppositories may affect the rectal absorption. Smaller particles may increase absorption (Tanabe et al. 1984). Decreased absorption has been observed with smaller particles when the transport through the fatty base is the rate-limiting step of absorption (Stuurman-Bieze et al. 1978; Rutten-Kingma, de Blaey, and Polderman 1979).

## Vaginal Delivery

The vaginal route of administration may be used for local and systemic action (Chien 1992b). However, it has been considered limited for systemic drug delivery due to the influence of the menstrual cycle on characteristics of the vaginal tissue and secretions (composition and volume) (Robinson, Narducci, and Ueda 1996). Advantages of vaginal administration include avoidance of gastrointestinal degradation and avoidance of hepatic first-pass metabolism. It may be useful for drugs with poor oral bioavailability or those that produce gastric irritation (Chien 1992b).

The vagina is a thin-walled, fibromuscular cavity that extends from the vulva to the cervix. It is approximately 6 to 8.5 cm in length. The vagina consists of three layers—an outer fibrous layer, a middle muscular layer, and an epithelial layer. The vaginal epithelial surface is kept moist by cervical secretions. The composition and volume of secretions varies with age, stage of menstrual cycle, and degree of sexual excitement (Li, Robinson, and Lee 1987). The vaginal pH is normally acidic (around pH 4–5) (Chien 1992b). Subtle changes occur to the vaginal mucosa during the menstrual cycle (Chien 1992a). The rich blood supply to the vagina empties into the iliac veins (Chien 1992b).

The barriers to drug absorption consist of an aqueous diffusion layer adjacent to the vaginal membrane. The rate-limiting step of absorption for drugs with high membrane permeability is diffusion through the aqueous layer, whereas the rate of absorption for drugs with low permeability is determined by the membrane permeability (Li, Robinson, and Lee 1987). Drug absorption occurs by passive diffusion and active mechanisms, via lipoidal and aqueous pore pathways. The rate and extent of drug absorption is affected by physiological conditions, such as rate of blood perfusion, volume and composition of vaginal secretions, physiological status of the vaginal mucosal layer, and cyclic variations (Robinson, Narducci, and Ueda 1996).

Preparations used for vaginal delivery include pessaries and controlled release systems, such as rings or microcapsules (Robinson, Narducci, and Ueda 1996; Chien 1992b; Li, Robinson, and Lee 1987).

## Conclusions

A variety of routes of administration (oral, parenteral, respiratory, nasal, transdermal and topical, ocular, otic, rectal, and vaginal) are discussed in this chapter. Particle properties play a key role in delivery, tissue targeting, residence time, and disposition for local or systemically acting agents. Clearly, the role of particulates in delivery and efficacy of drugs is of paramount importance. Characterizing the properties of particles and understanding the mechanisms of interaction with organs and tissues—physicochemically, physiologically, and pharmacologically—are significant steps in the development of novel dosage forms and effective drug delivery systems.

# 10

# General Conclusions

From the earliest use of powders alone to their use in capsules and tablets, the method of manufacture, physical form, and characteristics of a dosage form have been known to play key roles in the performance and ultimately the efficacy of drugs. Significant advances in the last 25 years in targeted systems to control drug delivery to the site of action or absorption have improved efficacy while minimizing toxicity. The therapeutic effect of drugs delivered in this manner, as compared to traditional dosage forms, is more highly dependent on the nature of the particulate system being employed, and sophisticated methods of analysis are required for its characterization. This book outlines the full complexity of pharmaceutical particulate systems and makes the case for giving consideration to all aspects of their performance, from manufacturing processes to therapeutic effect.

A classification framework is required in which to discuss the form of a particle before manufacturing can begin. The field of solid-state physics has already built the language by which particle structure can be discussed. The core element of this language is the crystal system defining the molecular array within a particle and the crystal habit, which describes the outward appearance of the particle following growth from a supersaturated medium. Particles can be manufactured in a multitude of ways, but each one of them will result to a greater or lesser degree in crystalline or amorphous particles. The form of the particles will dictate their physicochemical properties, which, in turn, through forces of interaction will dictate the state of the powder of which they are components.

Once particles have been constructed, and assessed as geometric forms, their dimensions must be described in general terms. Particle size and distribution form the basis for such descriptors. Particle size can be considered in terms of individual particles for which, again, common expressions must be employed, such as equivalent volume, surface, or projected area spheres. However, particles are found in distributions of sizes around a mean value. These distributions may be fitted to statistical distributions or mathematical functions, but it is important to recognize that such data models are only approximations of the real distribution. Therefore, the raw data may ultimately be the most valuable description of the size distribution. Particle size and distribution data may be employed to predict or understand phenomena ranging from dissolution properties to blending characteristics to bioavailability of the product.

Clearly, particle size and distribution are key properties from which many performance characteristics of a dosage form can be derived. However, before such correlations can occur, it is essential to select or devise an appropriate particle sizing technique. Of the range of methods available, some are direct, such as microscopy, whereas others are indirect, such as light scattering. It is imperative that the user understand the underlying theory for a method. Moreover, this understanding should be the basis for selecting appropriate methods for the application of the data. For example, Stokes' diameter, based on sedimentation, is most appropriate for suspension formulations. The complimentarity of particle sizing techniques should never be understated. Because each particle size descriptor and its method of analysis reflect a different facet of the same particle or population of particles, a great opportunity exists to use a multitude of techniques to build a picture of the powder.

At this juncture, the potential for error may seem to have been obviated by considering all of the issues above. Unfortunately, in routine activities many scientists erroneously come to this conclusion. The accuracy of any particle size determination depends most significantly on sampling. The entire population of particles can rarely be used as the sample, so the issue of representative sampling becomes very important. This sampling may be an essential part of whatever particle-sizing technique is employed, or it may be incumbent on the scientist to sample the powder prior to analysis and use. Fortunately, sampling has been studied in great detail and the statistical bases for good practices in sampling are established. Consequently, adopting good laboratory or manufacturing procedures based on sound engineering principles should allow the most accurate and precise data to be obtained.

The manufacture of the particle itself has been considered, but ultimately this will become a discrete element in the dosage form. It is frequently the case that drug particles are mixed with excipient particles to form homogeneous dispersions for a variety of purposes. It may then be necessary to examine the particulate properties of the blend, which in turn will require adequate sampling procedures.

Finally, the implications of the method of manufacture and characteristics must be considered from the perspective of in vitro testing for the quality of the product and in vivo performance for the efficacy and toxicity of the product. The fondest wish of pharmaceutical scientists is to correlate in vitro with in vivo performance and to approach the metaphorical Holy Grail of an in vitro–in vivo correlation (IVIVC). Although a great deal of research has been conducted in this area with some degree of success, it remains true that in vitro testing is, at best, a qualitative predictor of in vivo performance. Consequently, these issues must be considered separately. The in vitro performance of a fully characterized product may be considered in terms of a number of parameters of varying importance. For example, flow is important for vial filling, dissolution is important for oral absorption, and dispersion is important for delivery of aerosols. Examples of in vivo properties include lung deposition for aerosol products and oral bioavailability (which relates to in vivo dissolution).

Pharmaceutical scientists now have many techniques at their disposal that allow them to appreciate pharmaceutical particulate dispersed systems with the alacrity once felt for visualizing sand, gravel, stones, rocks, and boulders. Increasingly, particle characterization is recognized as a key to the quality and therapeutic performance of particulate medicines. The importance of particle properties in the action of drugs delivered from sophisticated targeted and controlled delivery systems has increased the rigor with which they are evaluated. Discussions of the benefits of crystal engineering are becoming more common, and the control of processes for the production of particles is of interest to regulatory and pharmacopeial groups. The complimentarity of particle sizing techniques in elucidating the nature of particle morphology may be equated to the allegory of the blind ones and the elephant (see Shah 1993). Depending on which feature is measured, different conclusions as to the nature of the object might be drawn. Therefore, complimentarity of techniques gives greater insight into the entire nature of the particle or powder. Nevertheless, some unfounded consternation remains on the part of novices that all techniques do not give the same result.

Integrating optimal sampling techniques and particle sizing methods with their process application by considering the nature of particle interactions in the context of the environment in which they exist forms the basis for good practice in pharmaceutical particulate science. By educating those new to these concepts, and embracing the experiences of other industries with respect to particulate matter, the prospects of increased understanding, improved medications (in terms of both quality and efficacy), and appropriate regulations seems assured.

Figure 10.1 illustrates the core functions in pharmaceutical particulate science and their relationship to materials in the molecular or solid state. Once a drug molecule has been discovered or isolated, it will be evaluated in a preliminary fashion for efficacy. Assuming that the potential has been established for the drug to be active in specific diseases, the drug is then subjected to a manufacturing process to prepare particles. These particles can then be characterized to establish specifications for the quality of the final dosage form. In vitro and in vivo performance can be evaluated to establish the potential for efficacy in the final product. The cycle is then complete because efficacy was the justification for initiating the product development. Note that some of the in vitro performance characteristics are used to establish product quality.

**Figure 10.1** Core functions in pharmaceutical particulate science.

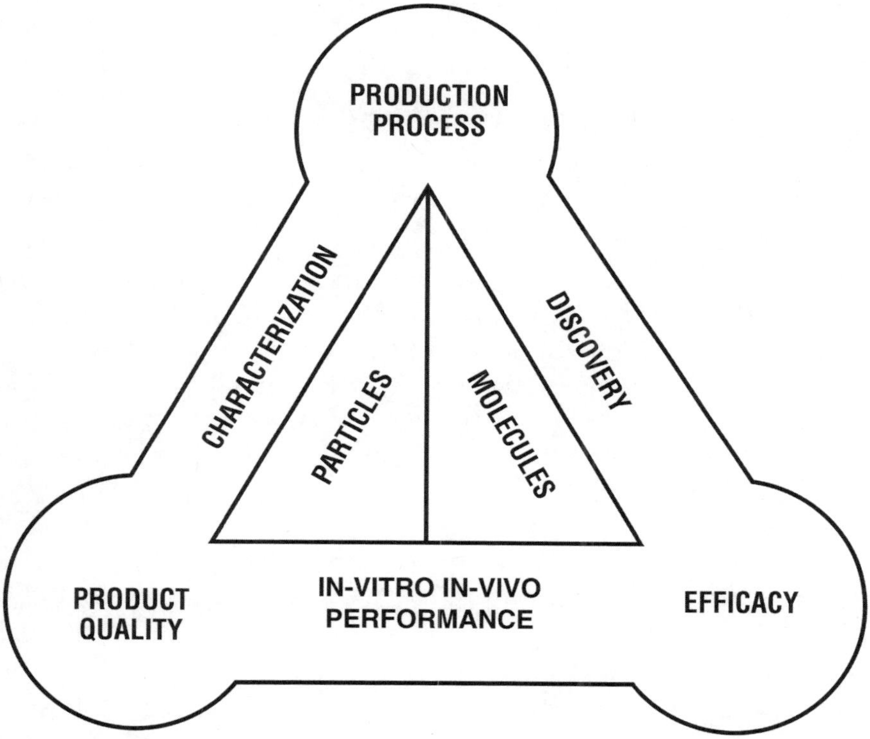


Clearly, particulate science plays a major role in the therapeutic application, and systematic studies are essential to the quality of the product. As the millennium evolves, greater understanding of all aspects of particle formation, morphology, degree of subdivision, and behavior will lead to increasingly sophisticated and therapeutically viable dosage forms that will ultimately revolutionize the delivery of drugs.

# References

Aguiar, A. J., J. E. Zelmer, and A. W. Kinkel. 1967. Deaggregation behavior of a relatively insoluble substituted benzoic acid and its sodium salt. *J. Pharm. Sci.* 56:1243–1252.

Alderborn, G. 1996. Particle dimensions. In *Pharmaceutical powder compaction technology*, eds. G. Alderborn and C. Nyström. New York: Marcel Dekker. 245–282.

Allen, L. V, V. A. Yanchick, and D. D. Maness. 1977. Dissolution rates of corticosteroids utilizing sugar glass dispersions. *J. Pharm. Sci.* 66:494–497.

Allen, T. 1968. *Particle size measurement*. London: Chapman and Hall.

Allen, T. 1975. Particle size, shape and distribution. In *Particle size measurement*, ed. T. Allen. London: Chapman and Hall. 74–112.

Allen, T. 1990a. *Particle size measurement*. London: Chapman and Hall.

Allen, T. 1990b. Sampling of dusty gases in gas streams. In *Particle size measurement*, ed. T. Allen. London: Chapman and Hall. 41–71.

Allen, T. 1990c. Sampling of powders. In *Particle size measurement*, ed. T. Allen. London: Chapman and Hall. 1–40.

Amidon, G. 1992. Report and recommendation of the USP Advisory Panel on physical test methods—functionality. I: General chapter on particle characterization by optical microscopy. *Pharm. Forum* 18:4089–4092.

Amidon, G. E. 1995. Physical and mechanical property characterization of powders. In *Physical characterization of pharmaceutical solids*, ed. H. G. Brittain. New York: Marcel Dekker. 281–319.

Anderberg, E. K., M. Bisrat, and C. Nyström. 1988. Physicochemical properties of drug release. Part 7: Effect of surfactant concentration and drug particle size on solubility and dissolution rate of felodipine, a sparingly soluble drug. *Int. J. Pharm.* 47:67–77.

Antal, E. J., C. F. Dick, C. E. Wright, I. R. Welshman, and E. M. Block. 1989. Comparative bioavailability of two medroxyprogesterone acetate suspensions. *Int. J. Pharm.* 54:33–39.

Aoyagi, N., H. Ogata, N. Kaniwa, M. Koibuchi, and T. Shibazaki. 1982a. Bioavailability of griseofulvin from tablets in beagle dogs and correlation with dissolution rate and bioavailability in humans. *J. Pharm. Sci.* 71:1169–1172.

Aoyagi, N., H. Ogata, N. Kaniwa, M. Koibuchi, and T. Shibazaki. 1982b. Bioavailability of griseofulvin from tablets in humans and correlation with dissolution rate. *J. Pharm. Sci.* 71:1165–1169.

ASTM (American Society for Testing and Materials). 1973. Recommended Practice for Analyses by Microscopial Methods. Philadelphia: ASTM.

Avnir, D., and M. Jaroniec. 1989. An isotherm equation for adsorption on fractal surfaces of heterogeneous porous materials. *Langmuir* 5:1431–1433.

Bapatia, K. M., and G. Hecht. 1996. Ophthalmic ointments and suspensions. In *Pharmaceutical dosage forms: Disperse systems,* eds. H. A. Lieberman, M. M. Rieger, and G. S. Banker. New York: Marcel Dekker. 357–397.

Barker, G., and M. Grimson. 1990. The physics of muesli. *New Scientist* 126:37–40.

Bastian, P., R. Bartkowski, H. Kohler, and T. Kissel. 1998. Chemo-embolization of experimental liver metastases. Part I: Distribution of biodegradable microspheres of different sizes in an animal model for the locoregional therapy. *Eur. J. Pharm. Biopharm.* 46:243–254.

Behl, C. R., H. K. Pimplaskar, A. P. Sileno, J. deMeireles, and V. D. Romeo. 1998. Effects of physicochemical properties and other factors on systemic nasal drug delivery. *Adv. Drug Deliv. Rev.* 29:89–116.

Bennema, P. 1992. Growth forms of crystals: Possible implications for powder technology. *Kona Powder Particle.* 10:25–40.

Bentejac, R. 1994. Freeze-drying. In *Powder technology and pharmaceutical processes,* eds. D. Chulia, M. Deleuil, and Y. Pourcelot. Amsterdam: Elsevier. 257–271.

Berglund, R. N., and B. Y. H. Liu. 1973. Generation of monodisperse aerosol standards. *Environ. Sci. Tech.* 7:147–153.

Berman, J., A. Schoeneman, and J. T. Shelton. 1996. Unit dose sampling: A tale of two thieves. *Drug Dev. Ind. Pharmacy* 22:1121–1132.

Berman, J., and J. A. Planchard. 1995. Blend uniformity and unit dose sampling. *Drug Dev. Ind. Pharmacy* 21:1257–1283.

Binnig, G., C. F. Quate, and C. Gerber. 1986. Atomic force microscope. *Phys. Rev. Lett.* 56:930–933.

Bisrat, M., M. Glazer, O. Camber, and P. Edman. 1992. Effect of particle size on ocular permeability of prednisolone acetate in rabbits. *Acta Pharm. Nordica* 4:5–8.

Boehme, G., H. Krupp, H. Rabenhorst, and G. Sandstede. 1962. Adhesion measurements involving small particles. *Trans. Inst. Chem. Engrs.* 40:252–259.

Bolton, S., 1997a. Analysis of variance. In *Pharmaceutical statistics: Practical and clinical applications,* ed. S. Bolton. New York: Marcel Dekker. 265–325.

Bolton, S., 1997b. Basic definitions and concepts. In *Pharmaceutical statistics: Practical and clinical applications,* ed. S. Bolton. New York: Marcel Dekker. 1–38.

Bolton, S., 1997c. Choosing samples. In *Pharmaceutical statistics: Practical and clinical applications,* ed. S. Bolton. New York: Marcel Dekker. 102–119.

Bolton, S, 1997d. Introduction to probability: The binomial and normal probability distributions. In *Pharmaceutical statistics: Practical and clinical applications*, ed. S. Bolton. New York: Marcel Dekker. 62–101.

Bolton, S, 1997e. Statistical inference: Estimation and hypothesis testing. In *Pharmaceutical statistics: Practical and clinical applications*, ed. S. Bolton. New York: Marcel Dekker. 120–188.

Booth, S. W., and J. M. Newton. 1987. Experimental investigation of adhesion between powders and surfaces. *J. Pharmacy Pharmacol.* 39:679–684.

*BP* (*British Pharmacopoeia*). 1999. General monographs on aerosols. London: The Stationary Office.

Brain, J. D., and J. D. Blanchard. 1993. Mechanisms of particle deposition and clearance. In *Aerosols in medicine, principles, diagnosis and therapy*, eds. F. Morén, M. B. Dolovich, M. T. Newhouse, and S. P. Newman. Amsterdam: Elsevier. 117–155.

Braun, M. A., R. Oschmann, and P. C. Schmidt. 1996. Influence of excipients and storage humidity on the deposition of disodium cromoglycate (DSCG) in the twin impinger. *Int. J. Pharm.* 135:53–62.

Bridgwater, J. 1994. Mixing. In *Powder technology and pharmaceutical processes*, eds. D. Chulia, M. Deleuil, and Y. Pourcelot. Amsterdam: Elsevier. 347–357.

Briggs, D., and M. P. Seah. 1990. *Practical surface analysis*. New York: Wiley.

Brittain, H. 1999. *Polymorphism in pharmaceutical solids*. New York: Marcel Dekker.

Brittain, H. 2000. Using single-crystal X-ray diffraction to study polymorphism and solvatochromism. *Pharm. Tech.* 24:116–125.

Brittain, H. 2001. Pharmaceutical applications of X-ray powder diffraction. *Pharm. Tech.* 25:142–150.

Brittain, H. G., S. J. Bogdanowich, D. E. Bugay, J. DeVincentis, G. Lewen, and A. W. Newman. 1991. Physical characterization of pharmaceutical solids. *Pharm. Res.* 8:963–973.

Brunauer, S., P. H. Emmett, and E. Teller. 1938. *J. Am. Chem. Soc.* 60:309.

Bryan, L., Y. Rungvejhavuttivittaya, and P. J. Stewart. 1979. Mixing and demixing of microdose quantities of sodium salicylate in a direct compression vehicle. *Powder Tech.* 22:147–151.

Buckton, G., and P. Darcy. 1999. Assessment of disorder in crystalline powders—a review of analytical techniques and their application. *Int. J. Pharm.* 179:141–158.

Burkoth, T. L., B. J. Bellhouse, G. Hewson, D. J. Longridge, A. G. Muddle, and D. F. Sarphie. 1999. Transdermal and transmucosal powdered drug delivery. *Crit. Rev. Ther. Drug Carrier Sys.* 16:331–384.

Butterstein, G. M., and V. D. Castracane. 2000. Effect of particle size on the prolonged action of subcutaneous danazol in male and female rats. *Fertil. Steril.* 74:356–358.

Byron, P. R. 1986. Some future perspective for unit dose inhalation aerosols. *Drug Dev. Ind. Pharmacy* 12:993–1015.

Byron, P. R., and A. J. Hickey. 1987. Spinning-disk generation and drying of monodisperse solid aerosols with output concentrations sufficient for single-breath inhalation studies. *J. Pharm. Sci.* 76:60–64.

Byron, P. R., J. Peart, and J. N. Staniforth. 1997. Aerosol electrostatics I: Properties of fine powders before and after aerosolization by dry powder inhalers. *Pharm. Res.* 14:698–705.

Caravati, C., F. Delogu, G. Cocco, and M. Rustici. 1999. Hyperchaotic qualities of the ball motion in a ball milling device. *Chaos* 9:219–226.

Carmichael, G. R. 1984. The effect of shape on particle solids flow. In *Particle characterization in technology*. Vol. 2: *Morphological analysis*, ed. J. K. Beddow. Boca Raton, Fla.: CRC Press. 205–221.

Carr, R. R. 1965. Evaluating flow properties of solids. *Chem. Eng*. Jan. 18:163–168.

Carstensen, J. 1993. *Pharmaceutical principles of solid dosage forms*. Lancaster, Pa.: Technomic.

Carstensen, J. T. 1980. Two-component systems. In *Solid pharmaceutics: Mechanical properties and rate phenomena*. New York: Academic Press. 102–134.

Carstensen, J. T., and M. V. Dali. 1996. Blending validation and content uniformity of low-content, noncohesive powder blends. *Drug Dev. Ind. Pharmacy* 22:285–290.

Castellanos, A., J. M. Valverde, A. T. Pérez, A. Ramos, and P. K. Watson. 1999. Flow regimes in fine cohesive powders. *Phys. Rev. Lett.* 82:1156–1159.

CFR (Code of Federal Regulations). 2000. Title 21, Part 211. Current good manufacturing practice for finished pharmaceuticals. Washington, D.C.: US Government Printing Office.

Chaumeil, J. C. 1998. Micronization: A method of improving the bioavailability of poorly soluble drugs. *Meth. Findings Exp. Clin. Pharmacol.* 20:211–215.

Chew, N., and H.-K. Chan. 1999. Influence of particle size, air flow and inhaler device on the dispersion of mannitol powders as aerosols. *Pharm. Res.* 16:1098–1103.

Chew, N. Y. K., D. F. Bagster, and H.-K. Chan. 2000. Effect of particle size, air flow and inhaler device on the aerosolisation of disodium cromoglycate powders. *Int. J. Pharm.* 206:75–83.

Chien, Y. W. 1992a. Mucosal drug delivery: Potential routes for noninvasive systemic administration. In *Novel drug delivery systems*, ed. Y. W. Chien. New York: Marcel Dekker. 197–228.

Chien, Y. W. 1992b. Oral drug delivery and delivery systems. In *Novel drug delivery systems*, ed. Y. W. Chien. New York: Marcel Dekker. 139–196.

Chien, Y. W. 1992c. Parenteral drug delivery and delivery systems. In *Novel drug delivery systems*, ed. Y. W. Chien. New York: Marcel Dekker. 381–528.

Chien, Y. W. 1992d. Vaginal drug delivery and delivery systems. In *Novel drug delivery systems*, ed. Y. W. Chien. New York: Marcel Dekker. 529–584.

Chien, Y. W., and S. F. Chang. 1985. Historic development of transnasal systemic medications. In *Transnasal systemic medications: Fundamental, developmental concepts and biomedical assessments*, ed. Y. W. Chien. Amsterdam: Elsevier. 1–99.

Chowhan, Z. T. 1995a. Segregation of particulate solids, Part I. *Pharm. Tech.* 5:56–70.

Chowhan, Z. T. 1995b. Segregation of particulate solids, Part II. *Pharm. Tech.* 6:80–94.

Claesson, P. M., T. Ederth, V. Bergeron, and M. W. Rutland. 1996. Techniques for measuring surface forces. *Adv. Colloid Interface Sci.* 67:119–183.

Clark, A. R. 1995. Medical aerosol inhalers: Past, present, and future. *Aerosol Sci. Tech.* 22:374–391.

Clark, A. R., and A. M. Hollingworth. 1993. The relationship between powder inhaler resistance and peak inspiratory conditions in healthy volunteers—implications for in vitro testing. *J. Aerosol Med.* 6:99–110.

Clarke, S. W., and S. P. Newman. 1984. Therapeutic aerosols. 2—Drugs available by the inhaled route. *Thorax* 39:1–7.

Colbeck, I. 1990. Dynamic shape factors of fractal clusters of carbonaceous smoke. *J. Aerosol Sci.* 21:S43–S46.

Concessio, N. M., M. M. van Oort, M. R. Knowles, and A. J. Hickey. 1999. Pharmaceutical dry powder aerosols: Correlation of powder properties with dose delivery and implications for pharmacodynamic effect. *Pharm. Res.* 16:828–834.

Concessio, N. M., M. M. van Oort, and A. J. Hickey. 1998. Impact force separation measurements their relevance in powder aerosol formulation. In *Respiratory drug delivery VI*, eds. R. N. Dalby, P. R. Byron, and S. J. Farr. Buffalo Grove, Ill.: Interpharm Press. 251–258.

Concessio, N. M. 1997. Dynamic properties of pharmaceutical powders and their impact on aerosol dispersion. Dissertation. University of North Carolina at Chapel Hill.

Conklin, J. D. 1978. The pharmacokinetics of nitrofurantoin and its related bioavailability. *Antibiot. Chemother.* 25:233–252.

Conklin, J. D., and F. J. Hailey. 1969. Urinary drug excretion in man during oral dosage of different nitrofurantoin formulations. *Clin. Pharmacol. Ther.* 10:534–539.

Corn, M. 1961a. The adhesion of solid particles to solid surfaces I: A review. *J. Air Poll. Control Assoc.* 11:523–528.

Corn, M. 1961b. The adhesion of solid particles to solid surfaces II. *J. Air Poll. Control Assoc.* 11:566–575.

Corn, M. 1966. Adhesion of particles. In *Aerosol science*, ed. C. N. Davies. New York: Academic Press. 359–392.

Cosslett, V. E., and W. C. Nixon. 1960. *X-ray microscopy*. Cambridge: Cambridge University Press.

Crooks, M. J., and R. Ho. 1976. Ordered mixing in direct compression of tablets. *Powder Tech.* 14:161–167.

Crowder, T. M., and A. J. Hickey. 1999. An instrument for rapid powder flow measurement and temporal fractal analysis. *Particle Particulate Sys. Char.*16:32–34.

Crowder, T. M., and A. J. Hickey. 2000. The physics of powder flow applied to pharmaceutical solids. *Pharm. Tech.* 24:50–58.

Crowder, T. M., V. Sethuraman, T. B. Fields, and A. J. Hickey. 1999. Signal processing and analysis applied to powder behaviour in a rotating drum. *Particle Particulate Sys. Char.* 16:191–196.

Crowder, T. M., M. D. Louey, V. V. Sethuraman, H. D. C. Smyth, and A. J. Hickey. 2001. 2001: An odyssey in inhaler formulations and design. *Pharm. Tech.* 25:99–113.

Crowder, T. M., J. A. Rosati, J. D. Schroeter, A. J. Hickey, and T. B. Martonen. 2002. Fundamental effects of particle morphology on lung delivery: Predictions of Stokes's law and the particular relevance to dry powder inhaler formulation and development. *Pharm. Res.* 19:239–245.

Dallavalle, J. M. 1948a. Characteristics of packings. In *Micromeritics*. New York: Pitman. 123–148.
Dallavalle, J. M. 1948b. Collection and separation of particulate matter from air. In *Micromeritics*. New York: Pitman. 435–349.
Dallavalle, J. M. 1948c. Determination of particle surface. In *Micromeritics*. New York: Pitman. 327–342.
Dallavalle, J. M. 1948d. Diffusion of particles. In *Micromeritics*. New York: Pitman. 164–179.
Dallavalle, J. M. 1948e. Introduction. In *Micromeritics*. New York: Pitman. 3–13.
Dallavalle, J. M. 1948f. Sampling. In *Micromeritics*. New York: Pitman. 479–495.
Dallavalle, J. M. 1948g. Shape and size-distribution of particles. In *Micromeritics*. New York: Pitman. 41–67.
Danilatos, G. D. 1991. Review and outline of environmental SEM at present. *J. Microscopy* 162:391.
Darcy, P., and G. Buckton. 1998. Crystallization of bulk samples of partially amorphous spray-dried lactose. *Pharm. Dev. Tech.* 3:503–507.
David, R., and D. Giron. 1994. Crystallization. In *Powder technology and pharmaceutical processes,* eds. D. Chulia, M. Deleuil, and Y. Pourcelot. Amsterdam: Elsevier. 193–242.
Davidson, J. A. 1995. Arrays of uniform latex spheres as test objects in optical microscopy. *Particle Particulate Sys. Char.* 12:173–178.
D'Emanuele, A., and C. Gilpin. 1996. Applications of the environmental scanning electron microscope to the analysis of pharmaceutical formulations. *Scanning* 18:522–527.
Deming, W. 1964. *Statistical adjustment of data*. Mineola, N.Y.: Dover.
Doelker, E. 1994. Assessment of powder compaction. In *Powder technology and pharmaceutical processes,* eds. D. Chulia, M. Deleuil, and Y. Pourcelot. Amsterdam: Elsevier. 403–471.
Dolovich, M. 1992. The relevance of aerosol particle size to clinical response. *J. Biopharm. Sci.* 3:139–145.
Donovan, M. D., and Y. Huang. 1998. Large molecule and particulate uptake in the nasal cavity: The effect of size on nasal absorption. *Adv. Drug Deliv. Rev.* 29:147–155.
Dorogovtsev, S. N. 1997. Avalanche mixing of granular materials: time taken for the disappearance of the pure fraction. *Tech. Phys.* 42:1346–1348.
Dressman, J. B., B. O. Palsson, A. Ozturk, and S. Ozturk. 1994. Mechanisms of release from coated pellets. In *Multiparticulate oral drug delivery,* ed. I. Ghebre-Sellassie. New York: Marcel Dekker. 285–306.
Ducker, W. A., T. J. Senden, and R. M. Pashley. 1991. Direct measurement of colloidal forces using an atomic force microscope. *Nature* 353:239–241.
Dunbar, C. A., and A. J. Hickey. 2000. Evaluation of probability density functions to approximate particle size distributions of representative pharmaceutical aerosols. *J. Aerosol Sci.* 31:813–831.
Dunbar, C., N. Concessio, and A. J. Hickey. 1998. Evaluation of atomizer performance in production of respirable spray-dried particles. *Pharm. Dev. Tech.* 3:433–441.
Dunbar, C. A., A. J. Hickey, and P. Holzner. 1998. Dispersion and characterization of pharmaceutical dry powder aerosols. *Kona* 16:7–45.

Duran, J. 2000. Ripples in tapped or blown powder. *Phys. Rev. Lett.* 84:5126–5129.

Edwards, C. 1997. Physiology of the colorectal barrier. *Adv. Drug Deliv. Rev.* 28:173–190.

Edwards, D. A., J. Hanes, G. Caponetti, J. Hrkach, A. Ben-Jebria, M. L. Eskew, J. Mintzes, D. Deaver, N. Lotan, and R. Langer. 1997. Large porous particles for pulmonary drug delivery. *Science* 276:1868–1871.

Edwards, D. A., A. Ben-Jebria, and R. Langer. 1998. Recent advances in pulmonary drug delivery using large, porous inhaled particles. *J. Appl. Physiol.* 85:379–385.

Esmen, N., and D. Johnson. 2001. Simulation analysis of inhalable dust sampling errors using multicomponent error model. *Aerosol Sci. Tech.* 35:824–828.

Esmen, N. A., and T. C. Lee. 1980. Distortion of cascade impactor measured size distribution due to bounce and blow-off. *Am. Indust. Hyg. Assoc. J.* 41:410–419.

Esposito, E., R. Roncarati, R. Cortesi, F. Cervellati, and C. Nastruzzi. 2000. Production of Eudragit microparticles by spray-drying technique: Influence of experimental parameters on morphological and dimensional characteristics. *Pharm. Dev. Tech.* 5:267–278.

Etzler, F. M., and R. Deanne. 1997. Particle size analysis: A comparison of various methods II. *Particle Particulate Sys. Char.* 14:278–282.

Etzler, F., and M. S. Sanderson. 1995. Particle size analysis: A comparative study of various methods. *Particle Particulate Sys. Char.* 12:217–224.

Evesque, P. 1991. Analysis of the statistics of sandpile avalanches using soil-mechanics results and concepts. *Phys. Rev. A* 43:2720–2740.

Fayed, M., and L. Otten. 1984. *Handbook of powder science and technology*. New York: Van Nostrand Reinhold.

Feldman, S. 1974. Physicochemical factors influencing drug absorption from the intramuscular injection site. *Bull. Parenteral Drug Assoc.* 28:53–63.

Ferguson, I. F. 1989. *Auger microprobe analysis*. New York: Adam Hilger.

Fernandez-Hervas, M. J., M.-A. Holgado, A.-M. Rabasco, and A. Fini. 1994. Use of fractal geometry on the characterization of particle morphology: Application to the diclofenac hydroxyethylpirrolidine salt. *Int. J. Pharm.* 108:187–194.

Fini, A., M. J. Fernandez-Hervas, M. A. Holgado, L. Rodriguez, C. Cavallari, N. Passerini, and O. Caputo. 1997. Fractal analysis of b-cyclodextrin-indomethacin particles compacted by ultrasound. *J. Pharm. Sci.* 86:1303–1309.

Firestone, B. A., M. A. Dickason, and T. Tran. 1998. Solubility characteristics of three fluoroquinolone ophthalmic solutions in an in vitro tear model. *Int. J. Pharm.* 164:119–128.

Flament, M. P., P. Leterme, and A. T. Gayot. 1995. Factors influencing nebulizing efficiency. *Drug Dev. Ind. Pharmacy* 21:2263–2285.

Floyd, A. G., and S. Jain. 1996. Injectable emulsions and suspensions. In *Pharmaceutical dosage forms: Disperse systems*, eds. H. A. Lieberman, M. M. Rieger, and G. S. Banker. New York: Marcel Dekker. 261–318.

Flynn, G. L. 1996. Cutaneous and transdermal delivery: Processes and systems of delivery. In *Modern pharmaceutics*, eds. G. S. Banker and C. T. Rhodes. New York: Marcel Dekker. 239–298.

Foldvari, M. 2000. Non-invasive administration of drugs through the skin: Challenges in delivery system design. *Pharm. Sci. Tech. Today* 3:417–425.

Foster, T. P., and M. W. Leatherman. 1995. Powder characteristics of proteins spray-dried from different spray-dryers. *Drug Dev. Ind. Pharmacy* 21:1705–1723.

Fuchs, N., and A. Sutugin. 1966. Generation and use of monodisperse aerosols. In *Aerosol science*, ed. C. Davies. London: Academic Press. 1–30.

Fuji, M., K. Machida, T. Takei, T. Watanabe, and M. Chikazawa. 1999. Effect of wettability on adhesion force between silica particles evaluated by atomic force microscopy measurement as a function of relative humidity. *Langmuir*. 15:4584–4589.

Fults, K. A., I. F. Miller, and A. J. Hickey. 1997. Effect of particle morphology on emitted dose of fatty acid-treated disodium cromoglycate powder aerosols. *Pharm. Dev. Tech*. 2:67–79.

Ganderton, D. 1992. The generation of respirable clouds from coarse powder aggregates. *J. Biopharm. Sci.* 3:101–105.

Garcia, T. P., M. K. Taylor, and G. S. Pande. 1998. Comparison of the performance of two sample thieves for the determination of the content uniformity of a powder blend. *Pharm. Dev. Tech*. 3:7–12.

Garekani, H., F. Sadeghi, A. Badiee, S. Mostafa, A. R. Rajabi-Siahboomi, and A. Rajabi-Siahboomi. 2001a. Crystal habit modifications of ibuprofen and their physicochemical characteristics. *Drug Dev. Ind. Pharmacy* 27:803–809.

Garekani, H., J. Ford, M. Rubinstein, and A. Rajabi-Siahboomi. 2001b. Effect of compression force, compression speed and particle size on the compression properties of paracetamol. *Drug Dev. Ind. Pharmacy* 27:935–942.

Geldart, D. 1986. Characterization of fluidized powders. In *Gas fluidization technology*. New York: Wiley. 33–51.

Gibbs, J. W. 1993. *The scientific papers of J. Willard Gibbs*. Woodbridge, Conn.: Ox Bow Press.

Giddings, C. J. 1993. Field-flow fractionation: analysis of macromolecular, colloidal and particulate materials. *Science* 260:1456–1465.

Glusker, J., M. Lewis, and M. Rossi. 1994. *Crystal structure analysis for chemists and biologists*. New York: VCH Publishers.

Goldman, A., and H. Lewis. 1984. Particle size analysis: Theory and statistical methods. In *Handbook of powder science and technology*, eds. M. Fayed and L. Otten. New York: Van Nostrand Reinhold. 1–30.

Goldstein, J. I., D. E. Newbury, P. Echlin, D. C. Joy, A. D. Romig, C. E. Lyman, C. Fiori, and E. Lifshin. 1992. *Scanning electron microscopy and x-ray microanalysis*. New York: Plenum Press.

Gonda, I. 1990. Aerosols for delivery of therapeutic and diagnostic agents to the respiratory tract. *Crit. Rev. Therapeut. Drug Carrier Sys.* 6:273–313.

Gonda, I. 1992. Targeting by deposition. In *Pharmaceutical inhalation aerosol technology*, ed. A. J. Hickey. New York: Marcel Dekker. 61–82.

Gonda, I., and P. R. Byron. 1978. Perspectives on the biopharmacy of inhalation aerosols. *Drug Dev. Ind. Pharmacy* 4:243–259.

Gonzalez-Rothi, R. J., and H. Schrier. 1995. Pulmonary delivery of liposome-encapsulated drugs in asthma therapy. *Clin. Immunotherapy* 4:331–337.

Grant, D. 1993. Report and recommendation of the USP Advisory Panel on physical test methods IV. Specific surface area determination of pharmaceutical powders by dynamic gas adsorption. *Pharm. Forum* 19:6171–6178.

Gray, P. 1964. *Handbook of basic microtechnique*. New York: McGraw-Hill.

Green, H., and W. Lane. 1964. Sampling and estimation. In *Particulate clouds, dusts, smokes and mists*. London: Spon Ltd. 227–283.

Greenblatt, D. J., T. W. Smith, and J. Koch-Weser. 1976. Bioavailability of drugs: The digoxin dilemma. *Clin. Pharmacokinetics* 1:36–51.

Guentensberger, J., P. Lameiro, A. Nyhuis, B. O'Connell, and A. Tigner. 1996. A statistical approach to blend uniformity acceptance criteria. *Drug Dev. Ind. Pharmacy* 22:1055–1061.

Gy, P. M. 1982a. Definition of basic terms and notations. In *Sampling of particulate materials: Theory and practice*. Amsterdam: Elsevier. 11–21.

Gy, P. M. 1982b. Increment delimitation error DE. In *Sampling of particulate materials: Theory and practice*. Amsterdam: Elsevier. 167–182.

Gy, P. M. 1982c. Increment extraction error EE. In *Sampling of particulate materials: Theory and practice*. Amsterdam: Elsevier. 183–212.

Gy, P. M. 1982d. Introduction. In *Sampling of particulate materials: Theory and practice*. Amsterdam: Elsevier. 1–9.

Gy, P. M. 1982e. Logical approach. In *Sampling of particulate materials: Theory and practice*. Amsterdam: Elsevier. 23–28.

Gy, P. M. 1982f. Periodic quality fluctuation error QE3. In *Sampling of particulate materials: Theory and practice*. Amsterdam: Elsevier. 113–120.

Gy, P. M. 1982g. Preparation errors PE. In *Sampling of particulate materials: Theory and practice*. Amsterdam: Elsevier. 325–332.

Gy, P. M. 1982h. Reference selection schemes. In *Sampling of particulate materials: Theory and practice*. Amsterdam: Elsevier. 75–83.

Gy, P. M. 1982i. Sampling processes. In *Sampling of particulate materials: Theory and practice*. Amsterdam: Elsevier. 33–38.

Gy, P. M. 1982j. Splitting methods and devices. In *Sampling of particulate materials: Theory and practice*. Amsterdam: Elsevier. 291–304.

Hadgraft, J. 2001. Skin, the final frontier. *Int. J. Pharm.* 224:1–18.

Heinz, W. F., and J. H. Hoh. 1999. Spatially resolved force spectroscopy of biological surfaces using the atomic force microscope. *Nanotechnology* 17:143–150.

Henein, H., J. K. Brimacombe, and A. P. Watkinson. 1983. Experimental study of transverse bed motion in rotary kilns. *Metallurg. Trans. B* 14:191–197.

Herdan, G., M. L. Smith, W. H. Hardwick, and P. Connor. 1960a. Microscope method of determining size distribution. In *Small particle statistics*. London: Butterworth-Heinemann. 329–337.

Herdan, G., M. L. Smith, W. H. Hardwick, and P. Connor. 1960b. Sampling procedures for particle size determinations. In *Small particle statistics*. London: Butterworth-Heinemann. 42–72.

Herdan, G., M. L. Smith, W. H. Hardwick, and P. Connor. 1960c. Sedimentation methods of determining particle size distribution. In *Small particle statistics*. London: Butterworth-Heinemann. 344–364.

Herdan, G., M. L. Smith, W. H. Hardwick, and P. Connor. 1960d. Statistical testing and analysis of differences between determination results. In *Small particle statistics*. London: Butterworth-Heinemann. 106–147.

Herdan, G., M. L. Smith, W. H. Hardwick, and P. Connor. 1960e. The units of particle statistics and the principles of particle size measurements. In *Small particle statistics*. London: Butterworth-Heinemann. 21–28.

Hersey J. A. 1975. Ordered mixing: A new concept in powder mixing practice. *Powder Tech.* 11:41–44.

Hersey, J. A. 1979. The development and applicability of powder mixing theory. *Int. J. Pharm. Tech. Prod. Manufac.* 1:6–13.

Hesketh, H. E. 1986a. Introduction. In *Fine particles in gaseous media*. Chelsea, U.K.: Lewis Publishers. 1–22.

Hesketh, H. E. 1986b. Unidirectional motion of particles. In *Fine particles in gaseous media*. Chelsea, U.K.: Lewis Publishers. 23–44.

Hickey, A. J. 1992. *Pharmaceutical inhalation aerosol technology*. New York: Marcel Dekker.

Hickey, A. J. 1996. *Inhalation aerosols: Physical and biological basis for therapy*. New York: Marcel Dekker.

Hickey, A. J., and N. M. Concessio. 1994. Flow properties of selected pharmaceutical powders from a vibrating spatula. *Particle Particulate Sys. Char* 11:457–462.

Hickey, A. J., and N. M. Concessio. 1997. Descriptors of irregular particle morphology and powder properties. *Adv. Drug Deliv. Rev.* 26:29–40.

Hickey, A. J., and C. A. Dunbar. 1997. A new millennium for inhaler technology. *Pharm. Tech.* 21:116–125.

Hickey, A. J., and D. Ganderton. 2001. *Pharmaceutical process engineering*. New York: Marcel Dekker.

Hickey, A. J., and T. B. Martonen. 1993. Behaviour of hygroscopic pharmaceutical aerosols and the influence of hydrophobic additives. *Pharm. Res.* 10:1–7.

Hickey, A. J., and D. C. Thompson. 1992. Physiology of the airways. In *Pharmaceutical inhalation aerosol technology*, ed. A. J. Hickey. New York: Marcel Dekker. 1–27.

Hickey, A. J., N. M. Concessio, M. M. van Oort, and R. M. Platz. 1994. Factors influencing the dispersion of dry powders as aerosols. *Pharm. Tech* 18:58–64.

Hilgraf, P. 1994. Pneumatic conveying. In *Powder technology and pharmaceutical processes*, eds. D. Chulia, M. Deleuil, and Y. Pourcelot. Amsterdam: Elsevier. 319–346.

Hindle, M., and P. R. Byron. 1995. Size distribution control of raw materials for dry-powder inhalers using the Aerosizer with the Aero-Disperser. *Pharm. Tech.* 19:64–78.

Hinds, W. C. 1982a. Acceleration and curvilinear particle motion. In *Aerosol technology: Properties, behavior and measurement of airborne particles*, W. C. Hinds. New York: Wiley. 104–126.

Hinds, W. C. 1982b. *Aerosol technology: Properties, behavior and measurement of airborne particles*. New York: Wiley.

Hinds, W. C. 1982c. Appendix. In *Aerosol technology: Properties, behavior and measurement of airborne particles*. New York: Wiley. 397–408.

Hinds, W. C. 1982d. Properties of gases. In *Aerosol technology: Properties, behavior and measurement of airborne particles*. New York: Wiley. 13–37.

Hinds, W. C. 1982e. Uniform particle motion. In *Aerosol technology: Properties, behavior and measurement of airborne particles*. New York: Wiley. 38–68.

Hinds, W. C. 1999. Uniform particle motion. In *Aerosol technology: Properties, behavior, and measurement of airborne particles*. New York: Wiley, 42–74.

Hirano, K., and H. Yamada. 1982. Studies on the absorption of practically water insoluble drugs following injection. Part 6: Subcutaneous absorption from aqueous suspensions in rats. *J. Pharm. Sci.* 71:500–505.

Hoener, B.-A., and L. Z. Benet. 1996. Factors influencing drug absorption and drug availability. In *Modern pharmaceutics*, eds. G. S. Banker and C. T. Rhodes. New York: Marcel Dekker. 121–153.

Hogan, J., P. I. Shue, F. Podczeck, and J. M. Newton. 1996. Investigations into the relationship between drug properties, filling, and the release of drugs from hard gelatin capsules using multivariate statistical analysis. *Pharm. Res.* 13:944–949.

Hornyak, G. L., S. T. Peschel, T. H. Sawitowski, and G. Schmid. 1998. TEM, STM and AFM as tools to study clusters and colloids. *Micron* 29:183–190.

Hundal, H. S., S. Rohani, H. C. Wood, and M. N. Pons. 1997. Particle shape characterization using image analysis and neural networks. *Powder Tech.* 91:217–227.

Hussain, A. A. 1998. Intranasal drug delivery. *Adv. Drug Deliv. Rev.* 29:39–49.

Hussain, N., V. Jaitley, and A. T. Florence. 2001. Recent advances in the understanding of uptake of microparticles across the gastrointestinal lymphatics. *Adv. Drug Deliv. Rev.* 50:107–142.

Hwang, R.-C., M. Gemoules, and D. Ramlose. 1998. A systematic approach for optimizing the blending process of a direct compression tablet formulation. *Pharm. Tech.* 22:158–170.

Israelachvili, J. N. 1991a. Adhesion. In *Intermolecular and surface forces*. London: Academic Press. 312–337.

Israelachvili, J. N. 1991b. Contrasts between intermolecular, interparticle and intersurface forces. In *Intermolecular and surface forces*. London: Academic Press. 152–175.

Israelachvili, J. 1991c. *Intermolecular and surface forces*. New York: Academic Press.

Jaeger, H. M., S. R. Nagel, and R. P. Behringer. 1996. The physics of granular materials. *Physics Today* 49:32–38.

Jaeghere, F. D., E. Allemann, J. Feijen, T. Kissel, E. Doelker, and R. Gurny. 2000. Freeze-drying and lyopreservative of diblock and triblock poly(lactic acid) – poly (ethylene oxide) (PLA-PEO) copolymer nanoparticles. *Pharm. Dev. Tech.* 5:473–483.

Jameel, F., K. Amberry, and M. Pikal. 2001. Freeze drying properties of some oligonucleotides. *Pharm. Dev. Tech.* 6:151–157.

Jani, P. U., G. W. Halbert, J. Langridge, and A. T. Florence. 1990. Nanoparticle uptake by the rat gastrointestinal mucosa: Quantitation and particle size dependency. *J. Pharmacy Pharmacol.* 42:821–826.

Jani, P. U., D. E. McCarthy, and A. T. Florence. 1992. Nanosphere and microsphere uptake via Peyer's patches: Observation of the rate of uptake in the rat after a single oral dose. *Int. J. Pharm.* 86:239–246.

Jarvinen, K., T. Jarvinen, and A. Urtti. 1995. Ocular delivery following topical delivery. *Adv. Drug Deliv. Rev.* 16:3–19.

Jenike, A. W. 1954. Flow of solids in bulk handling systems. *Bull. Univ. Utah* 64.

Jindal, K. C., R. S. Chaudry, A. K. Singal, S. S. Gangwal, and S. Khanna. 1995. Effect of particle size on the bioavailability and dissolution rate of rifampacin. *Indian Drugs* 32:100–115.

Jolliffe, I. G., and J. M. Newton. 1982. An investigation of the relationship between particle size and compression during capsule filling with an instrumented mG2 simulator. *J. Pharmacy Pharmacol.* 34:415–419.

Jolliffe, I. G., J. M. Newton, and J. K. Walters. 1980. Theoretical considerations of the filling of pharmaceutical hard gelatin capsules. *Powder Tech.* 27:189–195.

Jones, N. 2001. The nose and paranasal sinuses physiology and anatomy. *Adv. Drug Deliv. Rev.* 51:5–19.

Jones, W., and N. March. 1985a. *Theoretical solid state physics 1: Perfect lattices in equilibrium*. Mineola, N.Y.: Dover.

Jones, W., and N. March. 1985b. *Theoretical solid state physics 2: Non-equilibrium and disorder*. Mineola, N.Y.: Dover.

Kafui, K. D., and C. Thornton. 1997. Some observations on granular flow in hoppers and silos. In *Powders and grains '97*, eds. R. P. Behringer and J. T. Jenkins. Rotterdam, Netherlands: Balkema. 511–514.

Kano, J., A. Shimosaka, and J. Hidaka. 1997. A consideration of constitutive relationship between flowing particles. *Kona Powder Particle* 5:212–219.

Kasraian, K., T. Spitznagel, J. Juneau, and K. Yim. 1998. Characterization of the sucrose / glycine / water system by differential scanning calorimetry and freeze drying microscopy. *Pharm. Dev. Tech.* 3:233–239.

Kassem, N. M., K. K. L. Ho, and D. Ganderton. 1989. The effect of air flow and carrier size on the characteristics of an inspirable cloud. *J. Pharmacy Pharmacol.* 41:14P.

Kawashima, Y., T. Serigano, T. Hino, H. Yamamoto, and H. Takeuchi. 1998a. Design of inhalation dry powder of pranlukast hydrate to improve dispersibility by the surface modification with light anhydrous silicic acid (AEROSIL 200). *Int. J. Pharm.* 173:243–251.

Kawashima, Y., T. Serigano, T. Hino, H. Yamamoto, and H. Takeuchi. 1998b. A new powder design method to improve inhalation efficiency of pranlukast hydrate dry powder aerosols by surface modification with hydroxypropylmethylcellulose phthalate nanospheres. *Pharm. Res.* 15:1748–1752.

Kaye, B. H. 1981. *Direct characterization of fine particles*. New York: Wiley-Interscience.

Kaye, B. H. 1984. Fractal description of fine particle systems. In *Particle characterization in technology. Vol. 1: Applications and microanalysis*, ed. J. K. Beddow. Boca Raton, Fla.: CRC Press. 81–100.

Kaye, B. H. 1989. A random walk through fractal dimensions. New York: VCH Publishers.

Kaye, B. H. 1997. Characterizing the flowability of a powder using the concepts of fractal geometry and chaos theory. *Particle Particulate Sys. Char.* 14:53–66.

Kaye, B. H., D. Alliet, L. Switzer, and C. Turbitt-Daoust. 1999. Effect of shape on intermethod correlation of techniques for characterizing the size distribution of powder. Part 2: Correlating the size distribution as measured by diffractometer methods, TSI-Amherst aerosol spectrometer, and Coulter counter. *Particle Particulate Sys. Char.* 16:266–272.

Keskinen, J., K. Pietarinen, and M. Lehtimaki. 1992. Electrical low pressure impactor. *J. Aerosol Sci.* 23:353–360.

Khakhar, D. V., J. J. McCarthy, J. F. Gilchrist, and J. M. Ottino. 1999. Chaotic mixing of granular materials in two-dimensional tumbling mixers. *Chaos* 9:195–205.

Khati, S., and I. Shahrour. 1997. An experimental study of the evolution of density during flow of granular materials. In *Powders and grains '97*, eds. R. P. Behringer and J. T. Jenkins. Rotterdam, Netherlands: Balkema. 491–494.

Kirsch, L., S. Zhang, W. Muangsiri, M. Redmon, P. Luner, and D. Wurster. 2001. Development of a lyophilized formulation for (R,R)-formoterol (L)- tartrate. *Drug Dev. Ind. Pharmacy* 27:89–96.

Kordecki, M. C., and C. Orr. 1960. Adhesion of solid particles to solid surfaces. *Am. Med. Assoc. Arch. Environ. Health* 1:13–21.

Koval, M., K. Preiter, C. Adles, P. D. Stahl, and T. H. Steinberg. 1998. Size of IgG-opsonized particles determines macrophage response during internalization. *Exp. Cell Res.* 242:265–273.

Kublik, H., and M. T. Vidgren. 1998. Nasal delivery systems and their effect on deposition and absorption. *Adv. Drug Deliv. Rev.* 29:157–177.

Kulvanich, P., and P. J. Stewart. 1987a. Correlation between total adhesion and charge decay of a model interactive system during storage. *Int. J. Pharm.* 39: 51–57.

Kulvanich, P., and P. J. Stewart. 1987b. The effect of blending time on particle adhesion in a model interactive system. *J. Pharmacy Pharmacol.* 39:732–733.

Kulvanich, P., and P. J. Stewart. 1987c. The effect of particle size and concentration on the adhesive characteristics of a model drug-carrier interactive system. *J. Pharmacy Pharmacol.* 39:673–678.

Kulvanich, P., and P. J. Stewart. 1987d. An evaluation of the air stream Faraday cage in the electrostatic charge measurement of interactive drug systems. *Int. J. Pharm.* 36:243–252.

Kulvanich, P., and P. J. Stewart. 1987e. Fundamental considerations in the measurement of adhesional forces between particles using the centrifuge method. *Int. J. Pharm.* 35:111–120.

Kulvanich, P., and P. J. Stewart. 1988. Influence of relative humidity on the adhesive properties of a model interactive system. *J. Pharmacy Pharmacol.* 40:453–458.

Kumar, V., J. Kang, and T. Yang. 2001. Preparation and characterization of spray-dried oxidized cellulose microparticles. *Pharm. Dev. Tech.* 6:449–458.

Lam, K. K., and J. M. Newton. 1991. Investigation of applied compression on the adhesion of powders to a substrate surface. *Powder Tech.* 65:167–175.

Lam, K. K., and J. M. Newton. 1992a. Effect of temperature on particulate solid adhesion to a substrate surface. *Powder Tech.* 73:267–274.

Lam, K. K., and J. M. Newton. 1992b. Influence of particle size on the adhesion behaviour of powders, after application of an initial press-on force. *Powder Tech.* 73:117–125.

Lam, K. K., and J. M. Newton. 1993. The influence of the time of application on contact pressure on particle adhesion to a substrate surface. *Powder Tech.* 76:149–154.

Lamey, P. J., and M. A. O. Lewis. 1990. Buccal and sublingual delivery of drugs. In *Routes of drug administration*, eds. A. T. Florence and E. G. Salole. London: Wright. 30–47.

Lantz, R. 1981. Size reduction. In *Pharmaceutical dosage forms*, eds. H. A. Lieberman and L. Lachman. New York: Marcel Dekker. 77–152.

Lantz, R. J. Jr., and J. B. Schwartz. 1981. Mixing. In *Pharmaceutical dosage forms*, eds. H. A. Lieberman and L. Lachman. New York: Marcel Dekker. 1–53.

Lawless, P. 2001. High-resolution reconstruction method for presenting and manipulating particle histogram data. *Aerosol Sci. Tech.* 34:528–534.

Laycock, S., and J. N. Staniforth. 1983. Problems encountered in accurate determination of interparticulate forces in ordered mixes. *Part. Sci. Tech.* 1:259–268.

Laycock, S., and J. N. Staniforth. 1984. A method for determining interparticulate force in ordered mixes. *Labo-Pharma: Problems Tech.* 32:185–189.

LeBelle, M. J., S. J. Graham, E. D. Ormsby, R. M. Duhaime, R. C. Lawrence, and R. K. Pike. 1997. Metered-dose inhalers II: Particle size measurement variation. *Int. J. Pharm.* 151:209–221.

Leu, L.-P., J. T. Li, and C. M. Chen. 1997. Fluidization of group B particles in an acoustic field. *Powder Tech.* 94:23–28.

Lewis, A., and G. Simpkin. 1994. Tabletting—an industrial viewpoint. In *Powder technology and pharmaceutical processes*, eds. D. Chulia, M. Deleuil, and Y. Pourcelot. Amsterdam: Elsevier. 473–492.

Li, T., and K. Park. 1998. Fractal analysis of pharmaceutical particles by atomic force microscopy. *Pharm. Res.* 15:1222–1232.

Li, V. H. K., J. R. Robinson, and V. H. L. Lee. 1987. Influence of drug properties and routes of drug administration on the design of sustained and controlled release systems. In *Controlled drug delivery: Fundamentals and applications*, eds. J. R. Robinson and V. H. L. Lee. New York: Marcel Dekker. 3–94.

Liffman, K., G. Metcalfe, and P. Cleary. 1997. Convection due to horizontal shaking. In *Powders and grains '97*, eds. R. Behringer and J. T. Jenkins. Rotterdam, Netherlands: Balkema. 405–408.

Liversidge, G. G., and P. Conzentino. 1995. Drug particle size reduction for reducing gastric irritancy and enhancing absorptiion of naproxen in rats. *Int. J. Pharm.* 125:309–313.

Liversidge, G. G., and K. C. Cundy. 1995. Particle size reduction for improvement of oral bioavailability of hydrophobic drugs. Part 1: Absolute oral bioavailability of nanocrystalline danazol in beagle dogs. *Int. J. Pharm.* 125:91–97.

Llacer, J., V. Galardo, R. Delgado, J. Parraga, D. Martin, and M. Ruiz. 2001. X-ray diffraction and electron microscopy in the polymorphism study of Ondasetron hydrochloride. *Drug Dev. Ind. Pharmacy* 27:899–908.

Louey, M. D., P. Mulvaney, and P. J. Stewart. 2001. Characterization of adhesional properties of lactose carriers using atomic force microscopy. *J. Pharm. Biomed. Anal.* 25:559–567.

Lucas, P., K. Anderson, and J. N. Staniforth. 1998. Protein deposition from dry powder inhalers: fine particle multiplets as performance modifiers. *Pharm. Res.* 15:562–569.

Luerkens, D. 1991. *Theory and application of morphological analysis fine particles and surfaces*. Boca Raton, Fla.: CRC Press.

Luerkens, D. W., J. K. Beddow, and A. F. Vetter. 1984. Theory of morphological analysis. In *Particle characterization in technology. Vol 2. Morphological analysis,* ed. J. K. Beddow. Boca Raton, Fla.: CRC Press. 3–14.

Luner, P., S. Majuru, J. Seyer, and M. Kemper. 2000. Quantifying crystalline form composition in binary powder mixtures using near-infrared reflectance spectroscopy. *Pharm. Dev. Tech.* 5:231–246.

Ma, Z., H. G. Merkus, J. G. A. E. de Smet, C. Heffels, and B. Scarlett. 2000. New developments in particle characterization by laser diffraction: Size and shape. *Powder Tech.* 111:66–78.

Mackay, M., J. Phillips, and J. Hastewell. 1997. Peptide drug delivery: Colonic and rectal absorption. *Adv. Drug Deliv. Rev.* 28:253–273.

Mackenzie, R. 1970. *Differential thermal analysis*. London: Academic Press.

Mackin, L. A., G. Rowley, and E. J. Fletcher. 1997. An investigation of carrier particle type, electrostatic charge and relative humidity on in-vitro drug deposition from dry powder inhaler formulations. *Pharm. Sci.* 3:583–586.

Martin, A. 1993a. Coarse dispersions. In *Physical pharmacy*. Philadelphia: Lea and Febiger. 477–511.

Martin, A. 1993b. Colloids. In *Physical pharmacy*. Philadelphia: Lea and Febiger. 393–422.

Martin, A. 1993c. Diffusion and dissolution. In *Physical pharmacy*. Philadelphia: Lea and Febiger. 324–361.

Martin, A. 1993d. Interfacial phenomena. In *Physical pharmacy*. Philadelphia: Lea and Febiger. 362–392.

Martin, A. 1993e. Micromeritics. *Physical pharmacy* In *Physical pharmacy*. Philadelphia: Lea and Febiger. 423–452.

Martin, A. 1993f. *Physical pharmacy*. Philadelphia: Lea and Febiger.

Martin, A. 1993g. Rheology. In *Physical pharmacy*. Philadelphia: Lea and Febiger. 453–476.

Martonen, T. B., and I. M. Katz. 1993. Deposition patterns of aerosolized drugs within human lungs: Effects of ventilatory parameters. *Pharm. Res.* 10:871–878.

May, K. R. 1982. A personal note on the history of the cascade impactor. *J. Aerosol Sci.* 13:37–47.

Mayersohn, M. 1996. Principles of drug absorption. In *Modern pharmaceutics,* eds. G. S. Banker and C. T. Rhodes. New York: Marcel Dekker. 21–73.

Mbali-Pemba, C., and D. Chulia. 1995. Lactose texture monitoring during compacting. II: Detailed textural analysis in pore groups. *Drug Dev. Ind. Pharmacy* 21:623–631.

McCallion, O. N. M., K. M. G. Taylor, P. A. Bridges, M. Thomas, and A. J. Taylor. 1996. Jet nebulisers for pulmonary drug delivery. *Int. J. Pharm.* 130:1–11.

McGinity, J. W., C. T. Ku, R. Bodmeier, and M. R. Harris. 1985. Dissolution and uniformity properties of ordered mixes of micronized griseofulvin and a directly compressible excipient. *Drug Dev. Ind. Pharmacy* 11:891–900.

Melo, F., P. B. Umbanhowar, and H. L. Swinney. 1995. Hexagons, kinks, and disorder in oscillated granular layers. *Phys. Rev. Lett.* 75:3838–3841.

Menon, A., and N. Nerella. 2001. Effective graphical displays. *Pharm. Dev. Tech.* 6:477–484.

Menon, A., W. Drefko, and S. Chakrabarti. 1996. Comparative pharmaceutical evaluation of crospovidones. *Eur. J. Pharm. Sci.* 4:S186.

Metcalfe, G., T. Shinbrot, J. J. McCarthy, and J. M. Ottino. 1995. Avalanche mixing of granular solids. *Nature* 374:39–41.

Michoel, A., P. Rombaut, and A. Verhoye. 2002. Comparative evaluation of co-processed lactose and microcrystalline cellulose with their physical mixtures in the formulation of folic acid tablets. *Pharm. Dev. Tech.* 7:79–87.

Mitchell, J. P., P. A. Costa, and S. Waters. 1988. An assessment of an Andersen Mark-II cascade impactor. *J. Aerosol Sci.* 19:213–221.

Mizes, H. A. 1995. Surface roughness and particle adhesion. *J. Adhesion* 51:155–165.

Mohan, S., A. Rankell, C. Rehm, V. Bhalani, and A. Kulkarni. 1997. Unit-dose sampling and blend content uniformity testing. *Pharm. Tech.* 21:116–125.

Morén, F. 1993. Aerosol dosage forms and formulations. In *Aerosols in medicine, principles, diagnosis and therapy*, eds. F. Morén, M. B. Dolovich, M. T. Newhouse, and S. P. Newman. Amsterdam: Elsevier. 321–349.

Morgan, B. B. 1961. The adhesion and cohesion of fine particles. *Br. Coal Util. Res. Assoc. Monthly Bull.* 25:125–137.

Morgan, T. M., R. A. Parr, B. L. Reed, and B. C. Finnin. 1998. Enhanced transdermal delivery of sex hormones in swine with a noval topical aerosol. *J. Pharm. Sci.* 87:1219–1225.

Mori, S., T. Haruta, A. Yamamoto, I. Yamada, and E. Mizutani. 1991. Vibrofluidization of very fine particles. *Int. Chem. Engr.* 31:475–480.

Mosharraf, M., and C. Nystrom. 1995. Effect of particle size and shape on the surface specific dissolution rate of micronized practically insoluble drugs. *Int. J. Pharm.* 122:35–47.

Mullin, J. 1993. *Crystallization*. Oxford: Butterworth-Heinemann.

Mullins, M. E., L. P. Michaels, V. Menon, B. Locke, and M. B. Ranade. 1992. Effect of geometry on particle adhesion. *Aerosol Sci. Tech.* 17:105–118.

Mura, P., M. Faucci, and P. Parrini. 2001. Effects of grinding with microcrystalline cellulose and cyclodextrins on the ketoprofen physico-chemical properties. *Drug Dev. Ind. Pharmacy* 27:119–128.

Muzzio, F. J., P. Robinson, C. Wightman, and D. Brone. 1997. Sampling practices in powder blending. *Int. J. Pharm.* 155:153–178.

Muzzio, F. J., M. Roddy, D. Brone, A. W. Alexander, and O. Sudah. 1999. An improved powder-sampling tool. *Pharm. Tech.* 23:92–110.

Nabekura, T., Y. Ito, H. Cai, M. Terao, and R. Hori. 2000. Preparation and in vivo ocular absorption studies of disulfiram solid dispersion. *Biol. Pharm. Bull.* 23:616–620.

Nakagawa, M., S. A. Altobelli, A. Caprihan, and E. Fukushima. 1997. NMR Measurement and approximate derivation of the velocity depth profile of granular flow in a rotating partially filled, horizontal cylinder. In *Powders and grains '97*, eds. R. Behringer and J. Jenkins. Rotterdam, Netherlands: Balkema. 447–450.

Nash, R. A. 1996. Pharmaceutical suspensions. In *Pharmaceutical dosage forms. Disperse systems*, eds. H. A. Lieberman, M. M. Rieger, and G. S. Banker. New York: Marcel Dekker. 1–46.

Neumann, B. S. 1967. The flow properties of powders. In *Advances in pharmaceutical sciences*, eds. H. S. Bean, J. E. Carless, and A. H. Beckett. New York: Academic Press. 181–221.

Neuvonen, P. J. 1979. Bioavailability of phenytoin: Clinical pharmacokinetics and therapeutic implications. *Clin. Pharmacokinetics* 4:91–103.

Neuvonen, P. J., P. J. Pentikainen, and S. M. Elfving. 1977. Factors affecting the bioavailability of phenytoin. *Int. J. Clin. Pharmacol. Biopharmacy* 15:84–89.

Newhouse, M., and M. Dolovich. 1986. Aerosol therapy of asthma: Principles and applications. *Respiration* 50:123–130.

Newman, S. P. 1985. Aerosol deposition considerations in inhalation therapy. *Chest* 88:1525–1605.

Newman, S. P., and S. W. Clarke. 1983. Therapeutic aerosols 1—physical and practical considerations. *Thorax* 38:881–886.

Newman, S. P., K. P. Steed, S. J. Reader, G. Hooper, and B. Zierenberg. 1996. Efficient delivery to the lungs of flunisolide aerosol from a new portable hand-held multidose nebulizer. *J. Pharm. Sci.* 85:960–964.

Newton, J. M., and F. Bader. 1981. The prediction of the bulk densities of powder mixtures, and its relationship to the filling of hard gelatin capsules. *J. Pharmacy Pharmacol.* 33:621–626.

Nilsson, P., M. Westerberg, and C. Nyström. 1988. Physicochemical aspects of drug release. Part 5: The importance of surface coverage and compaction on drug dissolution from ordered mixtures. *Int. J. Pharm.* 45:111–125.

Nimmerfall, F., and J. Rosenthaler. 1980. Dependence of area under the curve on proquazone particle size and in vitro dissolution rate. *J. Pharm. Sci.* 69:605–607.

Niven, R. W. 1992. Modulated drug therapy with inhalation aerosols. In *Pharmaceutical inhalation aerosol technology*, ed. A. J. Hickey. New York: Marcel Dekker. 321–359.

Niven, R. W. 1993. Aerodynamic particle size testing using a time-of-flight aerosol beam spectrometer. *Pharm. Tech.* 17:72–78.

Norris, D. A., N. Puri, and P. J. Sinko. 1998. The effect of physical barriers and properties on the oral absorption of particulates. *Adv. Drug Deliv. Rev.* 34:135–154.

Nuffield, E. 1966. *X-ray diffraction methods*. New York: Wiley.

Nuijen, B., B. Nuijen, M. Bouma, H. Talsma, C. Manada, J. Jimeno, L. Lopez-Lazaro, A. Bult, and J. Beijnen. 2001. Development of a lyophilized parenteral pharmaceutical formulation of the investigational polypeptide marine anticancer agent Kahalalide F. *Drug Dev. Ind. Pharmacy* 27:767–780.

Nyström, C., and P. G. Karehill. 1996. The importance of intermolecular bonding forces and the concept of bonding surface area. In *Pharmaceutical powder compaction technology*, eds. G. Alderborn and C. Nyström. New York: Marcel Dekker. 17–53.

Nyström, C., and K. Malmqvist. 1980. Studies on direct compression of tablets I. The effect of particle size in mixing finely divided powders with granules. *Acta Pharm. Suecica* 17:282–287.

Nyström, C., and M. Westerberg. 1986. Use of ordered mixtures for improving the dissolution rate of low solubility compounds. *J. Pharmacy Pharmacol.* 38:161–165.

Ofner, C., M. R. L. Schnaare, and J. B. Schwartz. 1996. Oral aqueous suspensions. In *Pharmaceutical dosage forms: Disperse systems*, eds. H. A. Lieberman, M. M. Rieger, and G. S. Banker. New York: Marcel Dekker. 149–181.

Ogura, Y. 2001. Preface. Drug delivery to the posterior segments of the eye. *Adv. Drug Deliv. Rev.* 52:1–3.

Oie, S., and L. Z. Benet. 1996. The effect of route of administration and distribution on drug action. In *Modern pharmaceutics*, eds. G. S. Banker and C. T. Rhodes. New York: Marcel Dekker. 155–178.

Okada, J., Y. Matsuda, and F. Yoshinobu. 1969. Measurement of the adhesive force of pharmaceutical powders by the centrifuge method. *Yakugaku Zasshi.* 89:1539–1544.

Okonogi, S., S. Puttipipatkhachorn, and K. Yamamoto. 2001. Thermal behavior of ursodeoxycholic acid in urea: Identification of anomalous peak in the thermal analysis. *Drug Dev. Ind. Pharmacy* 27:819–823.

Ormós, Z. D. 1994. Granulation and coating. In *Powder technology and pharmaceutical processes*, eds. D. Chulia, M. Deleuil, and Y. Pourcelot. Amsterdam: Elsevier. 359–376.

Orr, N. 1981. Quality control and pharmaceutics of content uniformity of medicines containing potent drugs with special reference to tablets. In *Progress in the quality control of medicines*, eds. P. Deasy and R. Timoney. Amsterdam: Elsevier. 193–256.

Otsuka, A., K. Iida, K. Danjo, and H. Sunada. 1983. Measurements of the adhesive force between particles of powdered organic substances and a glass substrate by means of the impact separation method. I: Effect of temperature. *Chem. Pharm. Bull.* 31:4483–4488.

Otsuka, A., K. Iida, K. Danjo, and H. Sunada. 1988. Measurements of the adhesive force between particles of powdered materials and a glass substrate by means of the impact separation method. III: Effect of particle shape and surface asperity. *Chem. Pharm. Bull.* 36:741–749.

Padmadisastra, Y., R. A. Kennedy, and P. J. Stewart. 1994a. Influence of carrier moisture adsorption capacity on the degree of adhesion of interactive mixtures. *Int. J. Pharm.* 104:R1–R4.

Padmadisastra, Y., R. A. Kennedy, and P. J. Stewart. 1994b. Solid bridge formation in sulphonamide-Emdex interactive systems. *Int. J. Pharm.* 112:55–63.

Pak, H. K., and R. P. Behringer. 1993. Surface waves in vertically vibrated granular materials. *Phys. Rev. Lett.* 71:1832–1835.

Pak, H. K., E. Van Doorn, and R. P. Behringer. 1995. Effects of ambient gases on granular materials under vertical vibration. *Phys. Rev. Lett.* 74:4643–4646.

Palmieri, G., G. Bonacucina, P. Martino, and S. Martelli. 2001. Spray drying as a method for microparticle controlled release systems preparation: Advantages and limits. I. Water-soluble drugs. *Drug Dev. Ind. Pharmacy* 27:195–204.

Parkkali, S., V. P. Lehto, and E. Laine. 2000. Applying image analysis in the observation of recrystallization of amorphous cefadroxil. *Pharm. Dev. Tech.* 5:433–438.

Patton, J. S., J. Bukar, and S. Nagarajan. 1999. Inhaled insulin. *Adv. Drug Deliv. Rev.* 35:235–247.

Pezin, E., M. P. Flament, P. Leterme, and A. T. Gayot. 1996. Method of calibration of the twin impinger. *Eur. J. Pharm. Sci.* 4:S142.

Pham, H., P. Luo, F. Genin, and A. Dash. 2001. Synthesis and characterization of hydroxyapatite-ciprofloxacin delivery systems by precipitation and spray drying technique. *AAPS Pharm. Sci. Tech.* 3:Article 1.

Pietsch, W. 1991. *Size enlargement by agglomeration*. New York: Wiley.

Piscitelli, D., S. Bigora, C. Propst, S. Goskonda, P. Schwartz, L. Lesko, L. Augsburger, and D. Young. 1998. The impact of formulation and process changes on in vitro dissolution and the bioequivalence of piroxicam capsules. *Pharm. Dev. Tech.* 3:443–452.

Podczeck, F. 1997. A shape factor to assess the shape of particles using image analysis. *Powder Tech.* 93:47–53.

Podczeck, F. 1998. Fundamentals of adhesion of particles to surfaces. In *Particle-particle adhesion in pharmaceutical powder handling*. London: Imperial College Press. 1–80.

Podczeck, F., and J. M. Newton. 1995. Development of an ultracentrifuge technique to determine the adhesion and friction properties between particles and surfaces. *J. Pharm. Sci.* 84:1067–1071.

Podczeck, F., J. M. Newton, and M. B. James. 1994. Assessment of adhesion and autoadhesion forces between particles and surfaces. Part I: The investigation of autoadhesion phenomena of salmeterol xinafoate and lactose monohydrate particles using compacted powder surfaces. *J. Adhesion Sci. Tech.* 8:1459–1472.

Podczeck, F., J. M. Newton, and M. B. James. 1995a. Adhesion and autoadhesion measurements of micronized particles of pharmaceutical powders to compacted powder surfaces. *Chem. Pharm. Bull.* 43:1953–1957.

Podczeck, F., J. M. Newton, and M. B. James. 1995b. The adhesion strength of particles of salmeterol base and a series of salmeterol salts to compacted lactose monohydrate surfaces. *J. Adhesion Sci. Tech.* 9:1547–1558.

Podczeck, F., J. M. Newton, and M. B. James. 1995c. Assessment of adhesion and autoadhesion forces between particles and surfaces. Part II: The investigation of adhesion phenomena of salmeterol xinafoate and lactose monohydrate particles in particle-on-particle and particle-on-surface contact. *J. Adhesion Sci. Tech.* 9:475–486.

Podczeck, F., J. M. Newton, and M. B. James. 1996. The influence of constant and changing relative humidity of the air on the autoadhesion force between pharmaceutical powder particles. *Int. J. Pharm.* 145:221–229.

Podczeck, F., J. M. Newton, and M. B. James. 1997. Variations in the adhesion force between a drug and carrier particles as a result of changes in the relative humidity of the air. *Int. J. Pharm.* 149:151–160.

Porter, S. C., and I. Ghebre-Sellassie. 1994. Key factors in the development of modified-release pellets. In *Multiparticulate oral drug delivery*, ed. I. Ghebre-Sellassie. New York: Marcel Dekker. 217–284.

Pouliquen, O. Fingering instability in granular flows. 1997. In *Powders and grains '97*, eds. R. P. Behringer and J. T. Jenkins. Rotterdam, Netherlands: Balkema. 451–454.

Poux, M., P. Fayolle, J. Bertrand, D. Bridoux, and J. Bousquet. 1991. Powder mixing—some practical rules applied to agitated systems. *Powder Tech.* 68:213–234.

Povey, M. J. W. 2000. Particulate characterization by ultrasound. *Pharm. Sci. Tech. Today* 3:373–380.

Prime, D., P. J. Atkins, A. Slater, and B. Sumby. 1997. Review of dry powder inhalers. *Adv. Drug Deliv. Rev.* 26:51–58.

Ramachandran, C., and D. Fleisher. 2000. Transdermal delivery of drugs for the treatment of bone diseases. *Adv. Drug Deliv. Rev.* 42:197–223.

Ranucci, J. 1992. Dynamic plume-particle size analysis using laser diffraction. *Pharm. Tech.* 16:109–114.

Reimschuessel, A., J. Macur, and J. Marti. 1988. Microscopy. In *A guide to materials characterization and chemical analysis*, ed. J. Sibilia. New York: VCH Publishers. 137–166.

Reist, P. C. 1993a. *Aerosol science and technology*. New York: McGraw-Hill.

Reist, P. C. 1993b. Particle size distributions. *Aerosol science and technology*. New York: McGraw- Hill. 13–30.

Rey, L., and J. May. 1999. *Freeze-drying/lyophilization of pharmaceutical and biological products*. New York: Marcel Dekker.

Rhodes, C. T. 1989. Disperse systems. In *Modern pharmaceutics*, eds. G. S. Banker and C. T. Rhodes. New York: Marcel Dekker. 327–354.

Rhodes, M. 1998. Mixing and segregation. In *Introduction to particle technology*. ed. M. Rhodes. Chichester, U.K.: Wiley. 223–240.

Ridolfo, A. S., L. Thompkins, L. D. Bethtol, and R. H. Carmichael. 1979. Benoxaprofen, a new anti-inflammatory agent: Particle size effect on dissolution rate and oral absorption in humans. *J. Pharm. Sci.* 68:850–852.

Rietema, K. 1991a. *The dynamics of fine particles*. New York: Elsevier.

Rietema, K. 1991b. General introduction. In *The dynamics of fine particles*. New York: Elsevier. 1–13.

Rietema, K. 1991c. General introduction. In *The dynamics of fine powders*. New York: Elsevier. 1–18.

Rime, A.-F., D. Massuelle, F. Kubel, H.-R. Hagemann, and E. Doelker. 1997. Compressibility and compactibility of powdered polymers: Poly(vinyl chloride) powders. *Eur. J. Pharm. Biopharm.* 44:315–322.

Riviere, J. C. 1990. *Surface analytical techniques*. Oxford: Clarendon Press.

Robinson, D. H., W. A. Narducci, and C. T. Ueda. 1996. Drug delivery and administration. In *Pharmacotherapy: A pathophysiologic approach*, eds. J. T. DiPiro, R. L. Talbert, G. C. Yee, G. R. Matzke, B. G. Wells, and L. M. Posey. Stamford, Conn.: Appletone & Lange. 49–76.

Rogers, T., K. Johnston, and R. Williams. 1998. Solution-based particle formation of pharmaceutical powders by supercritical or compressed fluid $CO_2$ and cryogenic spray-freezing technologies. *Drug Dev. Ind. Pharmacy* 27:1003–1015.

Rudnic, E. M., and M. K. Kottke. 1996. Tablet dosage forms. In *Modern pharmaceutics*, eds. G. S. Banker and C. T. Rhodes. New York: Marcel Dekker. 333–394.

Rukeyser, M. 1942. *Willard Gibbs*. Woodbridge, Conn.: Ox Bow Press.

Rutten-Kingma, J. J., C. J. de Blaey, and J. Polderman. 1979. Biopharmaceutical studies of fatty suspension suppositories. Part 3: Influence of particle size and concentration on bioavailability of lithium sulphate in rats. *Int. J. Pharm.* 3:187–194.

Sacchetti, M. 2000. General equations for in situ salt screening of multibasic drugs in multiprotic acids. *Pharm Dev. Tech.* 5:579–582.

Sacchetti, M., and M. M. van Oort. 1996. Spray-drying and supercritical fluid particle generation techniques. In *Inhalation aerosols: Physical and biological basis for therapy*, ed. A. J. Hickey. New York: Marcel Dekker. 337–384.

Saettone, M. F., and L. Salminen. 1995. Ocular inserts for topical delivery. *Adv. Drug Deliv. Rev.* 16:95–106.

Schmidt, P. C., and C. J. W. Rubensdorfer. 1994. Evaluation of Ludipress as a "multipurpose excipient" for direct compression. Part I: Powder characteristics and tableting properties. *Drug Dev. Ind. Pharmacy* 20:2899–2925.

Schoenwald, R. D., and P. Stewart. 1980. Effect of particle size on ophthalmic bioavailability of dexamethasone suspensions in rabbits. *J. Pharm. Sci.* 69:391–394.

Schulze, D. 1994. Storage, feeding, proportioning. In *Powder technology and pharmaceutical processes*, eds. D. Chulia, M. Deleuil, and Y. Pourcelot. Amsterdam: Elsevier. 285–317.

Schuster, J., R. Rubsamen, P. Lloyd, and J. Lloyd. 1997. The AERx aerosol delivery system. *Pharm. Res.* 14:354–357.

Seyer, J., and P. Luner. 2001. Determination of indomethacin crystallinity in the presence of excipients using diffuse reflectance near-infrared spectroscopy. *Pharm. Dev. Tech.* 6:573–582.

Shah, I., 1993. The blind ones and the matter of the elephant. In *Tales of the dervishes*. London: Arkana. 25–26.

Shakesheff, K. M., M. C. Davies, C. J. Roberts, S. J. B. Tendler, and P. M. Williams. 1996. The role of scanning probe microscopy in drug delivery research. *Crit. Rev. Ther. Drug Carrier Sys.* 13:225–256.

Shewhart, W. 1986. *Statistical method from the viewpoint of quality control*. Mineola, N.Y.: Dover.

Sindel, U., and I. Zimmermann. 1998. Direct measurement of interparticulate forces in particulate solids using an atomic force microscope. In *Proceedings of World Congress on Particle Technology 3IchemE*.

Smith, D. L. O., and R. A. Lohnes. 1984. Behavior of bulk solids. In *Particle characterization in technology. Vol.1: Applications and Microanalysis*, ed. J. K. Beddow. Boca Raton, Fla.: CRC Press. 102–133.

Smith, G. L., R. A. Goulbourn, R. A. P. Burt, and D. H. Chatfield. 1977. Preliminary studies of absorption and excretion of benoxaprofen in man. *Br. J. Clin. Pharmacol.* 4:585–590.

Smith, P. L. 1997. Peptide delivery via the pulmonary route: A valid approach for local and systemic delivery. *J. Controlled Release* 46:99–106.

Snell, N. J. C., and D. Ganderton. 1999. Assessing lung deposition of inhaled medications: Consensus statement. Workshop of the British Association for Lung Research, Institute of Biology. London, UK. 17 April 1998. *Resp. Med.* 93:123–133.

Somorjai, G. A. 1972. *Principles of surface chemistry*. Englewood Cliffs, N.J.: Prentice-Hall. 1972.

Srichana, T., G. P. Martin, and C. Marriott. 1998. On the relationship between drug and carrier deposition from dry powder inhalers in vitro. *Int. J. Pharm.* 167:13–23.

Standish, N. 1985. Studies of size segregation in filling and emptying a hopper. *Powder Tech.* 45:43–56.

Staniforth, J. N. 1987. Order out of chaos. *J. Pharmacy Pharmacol.* 39:329–334.

Staniforth, J. N. 1988. Powder flow. In *Pharmaceutics: The science of dosage form design,* ed. M. E. Aulton. New York: Churchill Livingstone. 600–615.

Staniforth, J. N. 1996. Improvements in dry powder inhaler performance: Surface passivation effects. *Drug Deliv. Lungs* 7: 86–89.

Staniforth, J. N., and J. E. Rees. 1982a. Effect of vibration time, frequency and acceleration on drug content uniformity. *J. Pharmacy Pharmacol.* 34:700–706.

Staniforth, J. N., and J. E. Rees. 1982b. Electrostatic charge interactions in ordered powder mixes. *J. Pharmacy Pharmacol.* 34:69–76.

Staniforth, J. N., and J. E. Rees. 1983. Segregation of vibrated powder mixes containing different concentrations of fine potassium chloride and tablet excipients. *J. Pharmacy Pharmacol.* 35:549–554.

Staniforth, J. N., J. E. Rees, F. K. Lai, and J. A. Hersey. 1981. Determination of interparticulate forces in ordered powder mixes. *J. Pharmacy Pharmacol.* 33:485–490.

Staniforth, J. N., J. E. Rees, F. K. Lai, and J. A. Hersey. 1982. Interparticulate forces in binary and ternary ordered powder mixes. *J. Pharmacy Pharmacol.* 34:141–145.

Staniforth, J. N., H. A. Ahmed, and P. J. Lockwood. 1989. Quality assurance in pharmaceutical powder processing: Current developments. *Drug Dev. Ind. Pharmacy* 15:909–926.

Stanley-Wood, N., and R. Lines. 1992. *Particle size analysis.* Cambridge: Royal Society of Chemistry.

Stark, R. W., T. Drobek, M. Weth, J. Fricke, and W. M. Heckl. 1998. Determination of elastic properties of single aerogel powder particles with the AFM. *Ultramicroscopy* 75:161–169.

Stavchansky, S., and W. Gowan. 1984. Evaluation of the bioavailability of a solid dispersion of phenytoin in polyethylene glycol 6000 and a commercial phenytoin sodium capsule in the dog. *J. Pharm. Sci.* 73:733.

Steckel, H., and B. W. Müller. 1997. In vitro evaluation of dry powder inhalers II: influence of carrier particle size and concentration on in vitro deposition. *Int. J. Pharm.* 154:31–37.

Stein, S. W. 1999. Size distribution measurements of metered dose inhalers using Andersen Mark II cascade impactors. *Int. J. Pharm.* 186:43–52.

Stewart, P. and B. Alway. 1995. Aggregation during the dissolution of diazepam in interactive mixtures. *Particulate Sci. Tech.* 13:213–226.

Stewart, P. J. 1981. Influence of magnesium stearate on the homogeneity of a prednisolone-granule ordered mix. *Drug Dev. Ind. Pharmacy* 7:485–495.

Stewart, P. J. 1986. Particle interaction in pharmaceutical systems. *Pharmacy Int.* 7:146–149.

Stockham, J., and E. Fochtman. 1979. *Particle size analysis.* Ann Arbor, Mich.: Ann Arbor Science.

Stuurman-Bieze, A. G. G., F. Moolenaar, A. J. M. Schoonen, J. Visser, and T. Huizinga. 1978. Biopharmaceutics of rectal administration of drugs in man. Part 2. Influence of particle size on absorption rate and bioavailability. *Int. J. Pharm.* 1:337–347.

Suzuki, H., M. Ogawa, K. Hironaka, K. Ito, and H. Sunada. 2001. A nifedipine coground mixture with sodium deoxycholate. I: Colloidal particle formation and solid state analysis. *Drug Dev. Ind. Pharmacy* 27:943–949.

Swanson, P. D., F. J. Muzzio, A. Annapragada, and A. Adjei. 1996. Numerical analysis of motion and deposition of particles in cascade impactors. *Int. J. Pharm.* 142:33–51.

Tadayyan, A., and S. Rohani. 2000. Modeling and nonlinear control of continuous KCl-NaCl crystallizer. *Part. Sci. Tech.* 18:329–357.

Takeuchi, H., T. Yasuji, H. Yamamoto, and Y. Kawashima. 2000. Temperature- and moisture-induced crystallization of amorphous lactose in composite particles with sodium alginate prepared by spray-drying. *Pharm. Dev. Tech.* 5:355–363.

Tanabe, K., S. Itoh, S. Yoshida, Y. Furuichi, and A. Kamada. 1984. Effect of particle size on rectal absorption of aminopyrine in rabbits and its release from different suppository bases. *J. Pharm. Sci. Tech.* Japan. 44:155–161.

Taylor, J. K. 1990. *Statistical techniques for data analysis*. Boca Raton, Fla.: CRC Press.

Taylor, K. M. G., and O. N. M. McCallion. 1997. Ultrasonic nebulisers for pulmonary drug delivery. *Int. J. Pharm.* 153:93–104.

Thibert, R., M. Akbarieh, and R. Tawashi. 1988. Application of fractal dimension to the study of the surface ruggedness of granular solids and excipients. *J. Pharm. Sci.* 77:724–726.

Thompson, D. C. 1992. Pharmacology of therapeutic aerosols. In *Pharmaceutical inhalation aerosol technology*, ed. A. J. Hickey. New York: Marcel Dekker. 29–59.

Thompson, S. K. 1992a. Introduction. In *Sampling*. New York: Wiley. 1–7.

Thompson, S. K. 1992b. *Sampling*. New York: Wiley.

Thornburg, J., S. J. Cooper, and D. Leith. 1999. Counting efficiency of the API Aerosizer. *J. Aerosol Sci.* 30:479–488.

Tiwary, A. 2001. Modification of crystal habit and its sole in dosage form performance. *Drug Dev. Ind. Pharmacy* 27:699–709.

Tomas, J. 2001a. Assessment of mechanical properties of cohesive particulate solids. Part 1: Particle contact constitutive model. *Particulate Sci. Tech.* 19:95–110.

Tomas, J. 2001b. Assessment of mechanical properties of cohesive particulate solids. Part 2: Powder flow criteria. *Particulate Sci. Tech.* 19:111–129.

Tong, W.-Q., and G. Whitesell. 1998. In situ salt screening—a useful technique for discovery support and preformulation studies. *Pharm Dev. Tech.* 3:215–223.

Turi, E., Y. Khanna, and T. Taylo. 1988. Thermal analysis. In *A guide to materials characterization and chemical analysis*, ed. J. Sibilia. New York: VCH Publishers. 205–228.

Umprayn, K., A. Luengtummuen, C. Kitiyadisai, and T. Pornpiputsakul. 2001. Modification of crystal habit of ibuprofen using the phase partition technique: Effect of Aerosil and Tween 80 in binding solvent. *Drug Dev. Ind. Pharmacy* 27:1047–1056.

U.S. Food and Drug Administration. Department of Health and Human Services. Office of Generic Drugs. 1999. Draft Guidance for Industry ANDAs; Blend Uniformity Analysis. Washington, D.C.

USP (United States Pharmacopoeial Convention, Inc.). 2000. 24th Convention: *The national formulary, XXIV*. Rockville, Md.: USP.

Vaithiyalingam, S., V. Agarwal, I. Reddy, M. Ashraf, and M. Khan. 2001. Formulation development and stability evaluation. *Pharm. Tech.* 25:38–48.

Van Doorn, E., and R. P. Behringer. 1997. Dilation of a vibrated granular layer. *Europhysics Lett.* 40:387–392.

Van Drunen, M. A., B. Scarlett, and J. C. M. Marijnissen. 1994. Photon correlation spectroscopy for analysis of aerosols. *J. Aerosol Sci.* 25:347–348.

Vaughn, N. P. 1989. The Andersen impactor: calibration, wall losses and numerical simulation. *J. Aerosol Sci.* 20:67–90.

Vavia, P., and N. Adhage. 2000. Freeze-dried inclusion complexes of tolfenamic acid with beta cyclodextrins. *Pharm. Dev. Tech.* 5:571–574.

Venables, H., and J. Wells. 2001. Powder mixing. *Drug Dev. Ind. Pharmacy* 27:599–612.

Villalobos-Hernandez, J., and L. Villafuerte-Robles. 2001. Effect of carrier excipient and processing on stability of indorenate hydrochloride/excipient mixtures. *Pharm. Dev. Tech.* 6:551–561.

Visser, J. 1989. An invited review. Van der Waals and other cohesive forces affecting powder fluidization. *Powder Tech.* 58:1–10.

Visser, J. 1995. Particle adhesion and removal: A review. *Particulate Sci. Tech.* 13:169–196.

Ward, G. H., and R. K. Schultz. 1995. Process-induced crystallinity changes in albuterol sulfate and its effect on powder physical stability. *Pharm. Res.* 12:773–779.

Watanabe, A. 1997. Polarizing microscopy of crystalline drugs based on the crystal habit determination for the purpose of a rapid estimation of crystal habits, particle sizes and specific surface areas of small crystals. *Yakugaku Zasshi.* 117:771–785.

Westerberg, M., and C. Nyström. 1993. Physicochemical aspects of drug release. Part 18: Use of a surfactant and a disintegrant for improving drug dissolution rate from ordered mixtures. *Int. J. Pharm.* 3:142–147.

White, J. G. 1999. In situ determination of delavirdine mesylate particle size in solid oral dosage forms. *Pharm. Res.* 16:545–548.

Wilson, R. G., F. A. Stevie, and C. W. Magee. 1989. *Secondary ion mass spectrometry: A practical handbook for depth profiling and bulk impurity analysis*. New York: Wiley-Interscience.

Woodhead, P. J., S. R. Chapman, and J. M. Newton. 1983. The vibratory consolidation of particle size fractions of powders. *J. Pharmacy Pharmacol.* 35:621–626.

Wostheinrich, K., and P. Schmidt. 2001. Polymorphic changes of thiamine hydrochloride during granulation and tableting. *Drug Dev. Ind. Pharmacy* 27:481–489.

Wynn, E. J. W., and M. J. Hounslow. 1997. Coincidence correction for electrical-zone (Coulter counter) particle size analysers. *Powder Tech.* 93:163–175.

Yi, H., B. Mittal, V. Puri, F. Li, and C. Mancino. 2001. Measurement of bulk mechanical properties and modeling the load-response of Rootzone sands: Part 1: Round and angular monosize and binary mixtures. *Particulate Sci. Tech.* 19:145–173.

Yip, C. W., and J. A. Hersey. 1977. Ordered or random mixing: Choice of system and mixer. *Drug Dev. Ind. Pharmacy* 3:429–438.

Zeng, X. M., G. P. Martin, and C. Marriot. 1995. The controlled delivery of drugs to the lungs. *Int. J. Pharm.* 124:149–164.

Zeng, X. M., G. P. Martin, S.-K. Tee, and C. Marriott. 1998. The role of fine particle lactose on the dispersion and deaggregation of salbutamol sulphate in an air stream in vitro. *Int. J. Pharm.* 176:99–110.

Zeng, X. M., G. P. Martin, S.-K. Tee, A. A. Ghoush, and C. Marriott. 1999. Effects of particle size and adding sequence of fine lactose on the deposition of salbutamol sulphate from a dry powder formulation. *Int. J. Pharm.* 182:133–144.

Zeng, X. M., K. H. Pandal, and G. P. Martin. 2000. The influence of lactose carrier on the content homogeniety and dispersibility of beclomethasone dipropionate from dry powder aerosols. *Int. J. Pharm.* 197:41–52.

Zimon, A. D. 1963. Adhesion of solid particles to a plane surface 2: Influence of air humidity on adhesion. *Colloid J. USSR* (English translation) 25:265–268.

Zimon, A. D. 1982a. Adhesion in a gas medium: Sources of adhesion. In *Adhesion of dust and powder*, ed. A. D. Zimon. New York: Consultants Bureau. 93–144.

Zimon, A.D. 1982b. Adhesion of variously shaped particles to rough surfaces. In *Adhesion of dust and powder*, ed. A. D. Zimon. New York: Consultants Bureau. 145–172.

Zimon, A.D. 1982c. Detachment of adherent particles in an air stream. In *Adhesion of dust and powder*, ed. A. D. Zimon. New York: Consultants Bureau. 307–347.

Zimon, A. D. 1982d. Fundamental concepts of particle adhesion. In *Adhesion of dust and powder*, ed. A. D. Zimon. New York: Consultants Bureau. 1–30.

Zimon, A.D. 1982e. Methods for determining adhesive force. In *Adhesion of dust and powder*, ed. A. D. Zimon. New York: Consultants Bureau. 69–91.

Zimon, A. D., and T. S. Volkova. 1965. Effect of surface roughness on dust adhesion. *Colloid J. USSR* (English translation) 27:306–307.

Ziskind, G., M. Fichman, and C. Gutfinger. 1995. Resuspension of particulates from surfaces to turbulent flows—review and analysis. *J. Aerosol Sci.* 26:613–644.

Zuidema, J., F. A. J. M. Pieters, and G. S. M. J. E. Duchateau. 1988. Release and absorption rate aspects of intramuscularly injected pharmaceuticals. *Int. J. Pharm.* 47:1–12.

Zuidema, J., F. Kadir, H. A. C. Titulaer, and C. Oussoren. 1994. Release and absorption rates of intramuscularly and subcutaneously injected pharmaceuticals (II). *Int. J. Pharm.* 105:189–207.

# Index